NAVEDTRA 10624-A1
MINOR REVISION
April 1988
0502-LP-001-1580

**Naval Education and
Training Command**

Training Manual
(TRAMAN)

Basic Machines

Basic Machines
By Naval Education

Print ISBN 13: 978-1-4209-7102-6

Cover Image: a detail of "Vintage model of a classic car engine with focus on pistons used for education purposes", by Martin Bergsma / Shutterstock Images.

Please visit *www.digireads.com*

The terms training manual (TRAMAN) and nonresident training course (NRTC) are now the terms used to describe Navy nonresident training program materials. Specifically, a TRAMAN includes a rate training manual (RTM), officer text (OT), single subject training manual (SSTM), or modular single or multiple subject training manual (MODULE); and a NRTC includes nonresident career course (NRCC), officer correspondence course (OCC), enlisted correspondence course (ECC) or combination thereof.

PREFACE

Basic Machines is written as a reference for the enlisted men in the Navy whose duties require knowledge of the fundamentals of machinery.

Beginning with the simplest of machines—the lever—the book proceeds with the discussion of block and tackle, wheel and axle, inclined plane, screw and gears. It explains the concepts of work and power, and differentiates between the terms "force" and "pressure." The fundamentals of hydrostatic and hydraulic mechanisms are discussed in detail. The final chapters include several examples of the combination of simple mechanisms to make complex machines.

As one of several basic Navy Training Courses, this book was prepared by the Education and Training Support Service, Washington, D.C., for the Chief of Naval Personnel.

Revised 1988

THE UNITED STATES NAVY

GUARDIAN OF OUR COUNTRY

The United States Navy is responsible for maintaining control of the sea and is a ready force on watch at home and overseas, capable of strong action to preserve the peace or of instant offensive action to win in war.

It is upon the maintenance of this control that our country's glorious future depends; the United States Navy exists to make it so.

WE SERVE WITH HONOR

Tradition, valor, and victory are the Navy's heritage from the past. To these may be added dedication, discipline, and vigilance as the watchwords of the present and the future.

At home or on distant stations we serve with pride, confident in the respect of our country, our shipmates, and our families.

Our responsibilities sober us; our adversities strengthen us.

Service to God and Country is our special privilege. We serve with honor.

THE FUTURE OF THE NAVY

The Navy will always employ new weapons, new techniques, and greater power to protect and defend the United States on the sea, under the sea, and in the air.

Now and in the future, control of the sea gives the United States her greatest advantage for the maintenance of peace and for victory in war.

Mobility, surprise, dispersal, and offensive power are the keynotes of the new Navy. The roots of the Navy lie in a strong belief in the future, in continued dedication to our tasks, and in reflection on our heritage from the past.

Never have our opportunities and our responsibilities been greater.

CONTENTS

CHAPTER   Page

1.   Levers 1

2.   Block and Tackle 10

3.   The Wheel and Axle 16

4.   The Inclined Plane and Wedge................. 23

5.   The Screw 26

6.   Gears.................................... 30

7.   Work 39

8.   Power 46

9.   Force and Pressure 50

10.   Hydrostatic and Hydraulic Machines 58

11.   Machine Elements and Basic Mechanisms 69

12.   Deleted 87

13.   Internal Combustion Engine 106

14.   Power Trains 130

15.   Basic Computer Mechanisms 150

INDEX 158

CREDITS

CHAPTER 1

LEVERS

YOUR HELPERS

Ships have evolved through the ages from crude rafts to the huge complex cruisers and carriers of today's Navy. It was a long step from oars to sails, and another long step from sails to steam. With today's modern nuclear-powered ships another long step has been taken. Each step in the progress of shipbuilding has involved the use of more and more machines, until today's Navy personnel are specialists in operating and maintaining machinery. The Boatswain operates the winches to hoist cargo and the anchor; the personnel in the engine room operate pumps, valves, generators, and other machines to produce and control the ship's power; personnel in the weapons department operate shell hoist, and rammers; elevate and train the guns and missile launchers; the cooks operate mixers and can openers; personnel in the CB rates drive trucks, operate cranes, graders, and bulldozers. In fact it is safe to say every rate in the Navy uses machinery some time during the day's work.

Each machine used aboard ship has made the physical work load of the crew lighter. You don't walk the capstan to raise the anchor, or heave on a line to sling cargo aboard. Machines have taken over these jobs, and have simplified and made countless others easier. Machines are your friends. They have taken much of the backache and drudgery out of a sailor's life. Reading this book should help you recognize and understand the operations of many of the machines you see about you.

WHAT IS A MACHINE?

As you look about you, you probably see half a dozen machines that you don't recognize as such. Ordinarily you think of a machine as a complex device—a gasoline engine or a typewriter. They are machines, but so is a hammer, a screwdriver, a ship's wheel. A machine is any device that helps you to do work. It may help by changing the amount of the force or the speed of action. For example, a claw hammer is a machine—you can use it to apply a large force for pulling out a nail. A relatively small pull on the handle produces a much greater force at the claws.

We use machines to TRANSFORM energy. For example, a generator transforms mechanical energy into electrical energy. We use machines to TRANSFER energy from one place to another. For example, the connecting rods, crankshaft, drive shaft, and rear axle transfer energy from the automobile engine to the rear wheels.

Another use of machines is to MULTIPLY FORCE. We use a system of pulleys (a chain hoist for example) to lift a heavy load. The pulley system enables us to raise the load by exerting a force which is smaller than the weight of the load. We must exert this force over a greater distance than the height through which the load is raised; thus, the load moves more slowly than the chain on which we pull. A machine enables us to gain force, but only at the expense of speed.

Machines may also be used to MULTIPLY SPEED. The best example of this is the bicycle, by which we gain speed by exerting a greater force.

Machines are also used to CHANGE THE DIRECTION OF A FORCE. For example, the signalman's halyard enables one end of the line to exert an upward force on a signal flag as a downward force is exerted on the other end.

There are only six simple machines—the LEVER, the BLOCK, the WHEEL and AXLE, the INCLINED PLANE, the SCREW, and the GEAR. However, physicists recognize that there are only two basic principles in machines; namely, the lever and the inclined plane. The wheel and

1

axle, the block and tackle, and gears may be considered levers. The wedge and the screw use the principle of the inclined plane.

When you are familiar with the principles of these simple machines, you can readily understand the operation of complex machines. Complex machines are merely combinations of two or more simple machines.

THE LEVER

The simplest machine, and perhaps the one with which you are most familiar, is the LEVER. A seasaw is a familiar example of a lever in which one weight balances the other.

There are three basic parts which you will find in all levers; namely, the FULCRUM (F), a force or EFFORT (E), and a RESISTANCE (R). Look at the lever in figure 1-1. You see the pivotal point F (fulcrum); the EFFORT (E) which you apply at a distance A from the fulcrum; and a resistance (R) which acts at a distance a from the fulcrum. Distances A and a are the lever arms.

CLASSES OF LEVERS

The three classes of levers are shown in figure 1-2. The location of the fulcrum (the fixed or pivot point) with relation to the resistance (or weight) and the effort determines the lever class.

First-Class Levers

In the first-class lever (fig. 1-2A), the fulcrum is located between the effort and the resistance. As mentioned earlier, the seesaw is a good example of the first-class lever.

The amount of weight and the distance from the fulcrum can be varied to suit the need. Another good example is the oars in a rowboat. Notice that the sailor in figure 1-3 applies his effort on the handles of the oars. The oarlock acts as the fulcrum, and the water acts as the resistance to be overcome. In this case, as in figure 1-1, the force is applied on one side of the fulcrum and the resistance to be overcome is applied to the opposite side, hence this is a first-class lever. Crowbars, shears, and pliers are common examples of this class of lever.

Second-Class Levers

The second-class lever (fig. 1-2B) has the fulcrum at one end; the effort is applied at the other end. The resistance is somewhere between these points. The wheelbarrow in figure 1-4 is a good example of a second-class lever. If you apply 50 pounds of effort to the handles of a wheelbarrow 4 feet from the fulcrum (wheel), you can lift 200 pounds of weight 1 foot from the fulcrum. If the load were placed farther back away from the wheel, would it be easier or harder to lift?

Both first- and second-class levers are commonly used to help in overcoming big resistances with a relatively small effort.

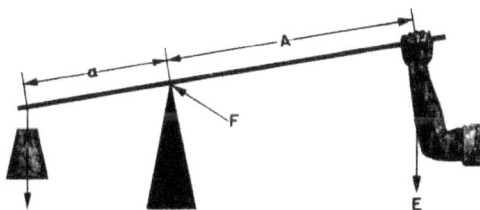

131.1

Figure 1-1.—A simple lever.

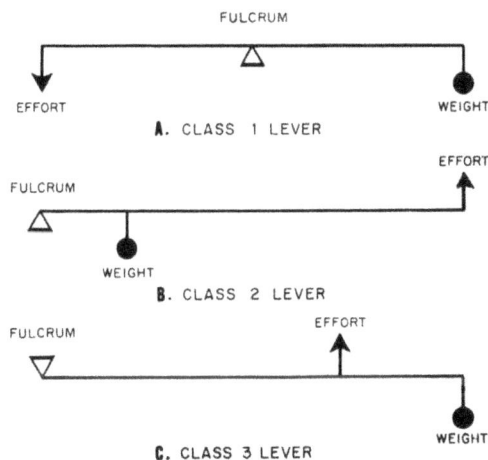

FULCRUM

EFFORT   WEIGHT

A. CLASS 1 LEVER

EFFORT

FULCRUM

WEIGHT

B. CLASS 2 LEVER

EFFORT

FULCRUM

WEIGHT

C. CLASS 3 LEVER

5.30

Figure 1-2.—Three classes of levers.

2

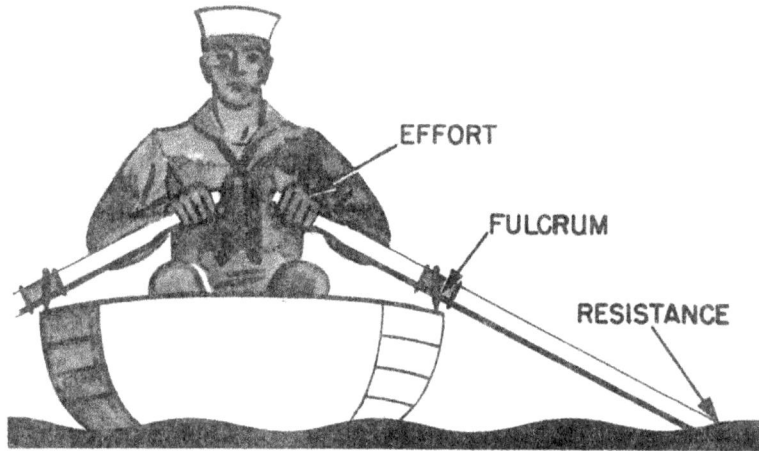

131.2

Figure 1-3.—Oars are levers.

Third-Class Levers

There are occasions when you will want to speed up the movement of the resistance even though you have to use a large amount of effort. Levers that help you accomplish this are third-class levers. As shown in figure 1-2C, the fulcrum is at one end of the lever and the weight or resistance to be overcome is at the

other end, with the effort applied at some point between. You can always spot third-class levers because you will find the effort applied between the fulcrum and the resistance. Look at figure 1-5. It is easy to see that while point E is moving the short distance e, the resistance R has been moved a greater distance r. The speed of R must have been greater than that of E, since R covered a greater distance in the same length of time.

Your arm (fig. 1-6), is a third-class lever. It is this lever action that makes it possible for you to flex your arms so quickly. Your elbow is the fulcrum. Your biceps muscle,

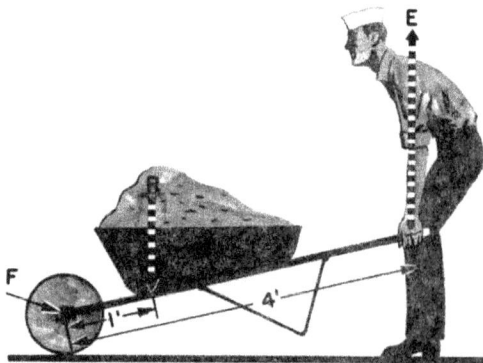

131.3

Figure 1-4.—This makes it easier.

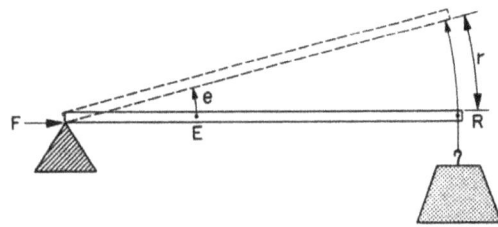

131.4

Figure 1-5.—A third-class lever.

3

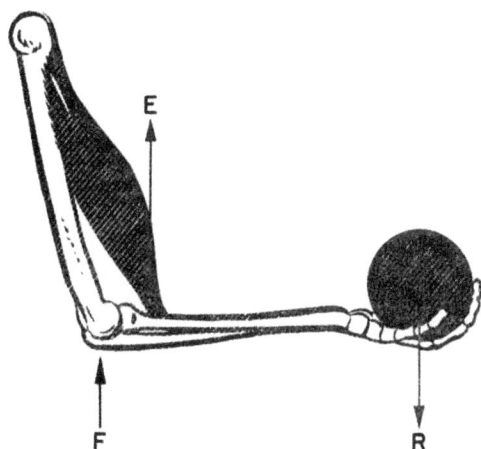

131.5
Figure 1-6.—Your arm is a lever.

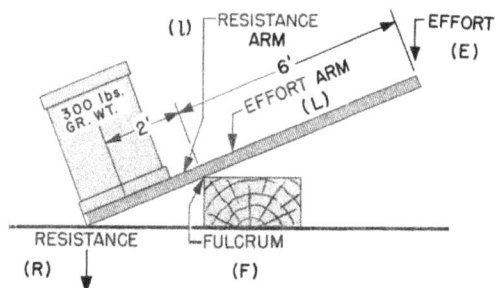

110.4
Figure 1-7.—Easy does it.

which ties onto your forearm about an inch below the elbow, applies the effort; and your hand is the resistance, located some 18 inches from the fulcrum. In the split second it takes your biceps muscle to contract an inch, your hand has moved through an 18-inch arc. You know from experience that it takes a big pull at E to overcome a relatively small resistance at R. Just to remind yourself of this principle, try closing a door by pushing on it about three or four inches from the hinges (fulcrum). The moral is, you don't use third-class levers to do heavy jobs, you use them to gain speed.

One convenient thing about machines is that you can determine in advance the forces required for their operation, as well as the forces they will exert. Consider for a moment the first-class lever. Suppose you have an iron bar, like the one shown in figure 1-7. This bar is 9 feet long, and you want to use it to raise a 300-pound crate off the deck while you slide a dolly under the crate. But you can exert only 100 pounds to lift the crate. So you place the fulcrum—a wooden block—beneath one end of the bar, and force that end of the bar under the crate. Then you push down on the other end of the bar. After a few adjustments of the position of the fulcrum, you will find that your 100-pound force will just fit the crate when the fulcrum is 2 feet from center of the crate.

This leaves a 6-foot length of bar from the fulcrum to the point where you push down. The 6-foot portion is three times as long as the distance from the fulcrum to the center of the crate. But you lifted a load three times as great as the force you applied—3 x 100 = 300 pounds. Here is an indication of a direct relationship between lengths of lever arms and forces acting on those arms.

You can state this relationship in general terms by saying—the length of the effort arm is the same number of times greater than the length of the resistance arm as the resistance to be overcome is greater than the effort you must apply. Writing these words as a mathematical equation, it looks like this—

$$\frac{L}{l} = \frac{R}{E}$$

in which,

L = length of effort arm.
l = length of resistance arm.
R = resistance weight or force.
E = effort force.

Remember that all distances must be in the same units—such as feet, and all forces must be in the same units—such as pounds.

Now take another problem and see how it works out. Suppose you want to pry up the lid of a paint can (fig. 1-8) with a 6-inch file scraper, and you know that the average force holding the lid is 50 pounds. If the distance from the edge of the paint can to the edge of

4

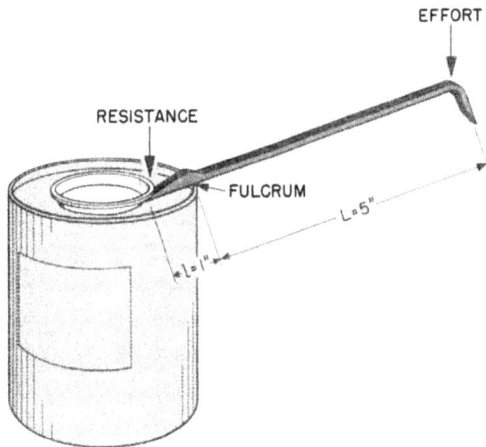

131.6
Figure 1-8.—A first-class job.

the cover is one inch, what force will you have to apply on the end of the file scraper?

According to the formula:

$$\frac{L}{1} = \frac{R}{E}$$

Here L = 5 inches; 1 = 1 inch; R = 50 pounds, and E is unknown.

Substitute the numbers in their proper places, Then,

$$\frac{5}{1} = \frac{50}{E}$$

and

$$E = \frac{50 \times 1}{5} = 10 \text{ pounds}$$

You will need to apply a force of only 10 pounds.

The same general formula applies for second-class levers. But you must be careful to measure the proper lengths of the effort arm and the resistance arm. Looking back at the wheelbarrow problem, assume that the length of the handles from the axle of the wheel—which is the fulcrum—to the grip is 4 feet. How long is the effort arm? You're right, it's 4 feet. If the center of the load of sand is 1 foot from

the axle, then the length of the resistance arm is 1 foot.

By substituting in the formula,

$$\frac{L}{1} = \frac{R}{E}$$

$$\frac{4}{1} = \frac{200}{E}$$

and

$$E = 50 \text{ lb.}$$

Now for the third-class lever. With one hand, you lift a projectile weighing approximately 10 pounds. If your biceps muscle attaches to your forearm 1 inch below your elbow, and the distance from the elbow to the palm of your hand is 18 inches, what pull must your muscle exert in order to hold the projectile and flex your arm at the elbow?

By substituting in the formula,

$$\frac{L}{1} = \frac{R}{E}, \text{ it becomes } \frac{1}{18} = \frac{10}{E}$$

and

$$E = 18 \times 10 = 180 \text{ lb.}$$

Your muscle must exert a 180-pound pull to hold up a 10-pound shell. Our muscles are poorly arranged for lifting or pulling—and that's why some work seems pretty tough. But remember, third-class levers are used primarily to speed up the motion of the resistance.

Curved Lever Arms

Up to this point you have been looking at levers with straight arms. In every case, the direction in which the resistance acts is parallel to the direction in which the effort is exerted. However, all levers are not straight. You'll need to learn to recognize all types of levers, and to understand their operation.

Look at figure 1-9. You may wonder how to measure the length of the effort arm, which is represented by the curved pump handle. You do not measure around the curve—you still use a straight-line distance. To determine the length of the effort arm, draw a straight line AB through the point where the effort is applied and in the direction that it is applied. From point E on this line, draw a second line EF that passes through the fulcrum and is perpendicular to line AB. The length of the line EF is the actual length L of the effort arm.

5

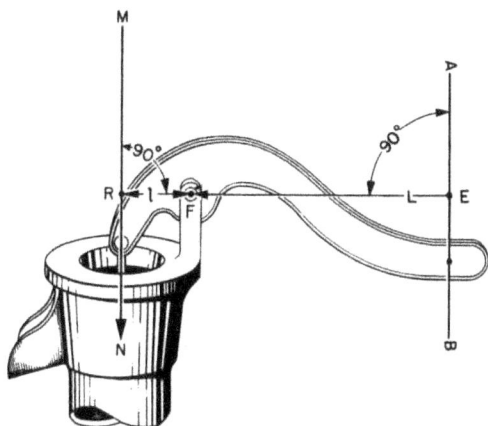

131.7

Figure 1-9.—A curved lever arm.

To find the length of the resistance arm, use the same method. Draw a line MN in the direction that the resistance is operating, and through the point where the resistance is attached to the other end of the handle. From point R on this line, draw a line RF perpendicular to MN so that it passes through the fulcrum. The length of RF is the length 1 of the resistance arm.

Regardless of the curvature of the handle, this method can be used to find the lengths L and 1. Then curved levers are solved just like straight levers.

MECHANICAL ADVANTAGE

There is another thing about first-class and second-class levers that you have probably noticed by now. Since they can be used to magnify the applied force, they provide positive mechanical advantages. The third-class lever provides what's called a fractional mechanical advantage, which is really a mechanical disadvantage—you use more force than the force of the load you lift.

In the wheelbarrow problem, you saw that a 50-pound pull actually overcame the 200-pound weight of the sand. The sailor's effort was magnified four times, so you may say that the mechanical advantage of the wheelbarrow is

4. Expressing the same idea in mathematical terms,

$$\text{MECHANICAL ADVANTAGE} = \frac{\text{RESISTANCE}}{\text{EFFORT}}$$

or

$$\text{M. A.} = \frac{R}{E}$$

Thus, in the case of the wheelbarrow,

$$\text{M.A.} = \frac{200}{50} = 4$$

This rule applies to all machines.

Mechanical advantage of levers may also be found by dividing the length of the effort arm A by the length of the resistance arm a. Stated as a formula, this reads:

$$\text{MECHANICAL ADVANTAGE} = \frac{\text{EFFORT ARM}}{\text{RESISTANCE ARM}}$$

$$\text{M. A.} = \frac{A}{a}$$

How does this apply to third-class levers? Your muscle pulls with a force of 1,800 pounds in order to lift a 100-pound projectile. So you have a mechanical advantage of $\frac{100}{1,800}$, or $\frac{1}{18}$, which is fractional—less than 1.

SUMMARY

Now for a brief summary of levers.

Levers are machines because they help you to do your work. They help by changing the size, direction, or speed of the force you apply.

There are three classes of levers. They differ primarily in the relative points where effort is applied, where the resistance is overcome, and where the fulcrum is located.

First-class levers have the effort and the resistance on opposite sides of the fulcrum, and effort and resistance move in opposite directions.

Second-class levers have the effort and the resistance on the same side of the fulcrum, but the effort is farther from the fulcrum than is the resistance. Both effort and resistance move in the same direction.

Third-class levers have the effort applied on the same side of the fulcrum as the resistance, but the effort is applied between the resistance and the fulcrum. Both move in the same direction.

First- and second-class levers can be used to magnify the amount of the effort exerted, and to decrease the speed of effort. First-class and third-class levers can be used to magnify the distance and the speed of the effort exerted, and to decrease its magnitude.

The same general formula applies to all three types of levers:

$$\frac{L}{l} = \frac{R}{E}$$

MECHANICAL ADVANTAGE (M.A.) is an expression of the ratio of the applied force and the resistance. It may be written:

$$M. A. = \frac{R}{E}$$

APPLICATIONS AFLOAT
AND ASHORE

Doors aboard a ship are locked shut by lugs called dogs. Figure 1-10 shows you how these dogs are used to secure the door. If the handle is four times as long as the lug, that 50-pound heave of yours is multiplied to 200 pounds against the slanting face of the wedge. Incidentally, take a look at that wedge—it's an inclined plane, and it multiplies the 200-pound

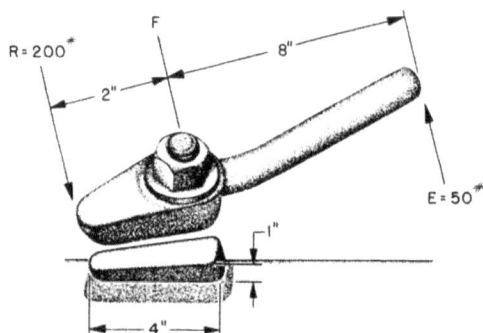

Figure 1 10.—It's a dog.

3.100

force by about four. Result—your 50-pound heave actually ends up as an 800-pound force on each wedge to keep the hatch closed! The hatch dog is one use of a first-class lever, in combination with an inclined plane.

The breech of a big gun is closed with a breech plug. Figure 1-11 shows you that this plug has some interrupted screw threads on it which fit into similar interrupted threads in the breech. Turning the plug part way around locks it into the breech. The plug is locked and unlocked by the operating lever. Notice that the connecting rod is secured to the operating lever a few inches from the fulcrum. You'll see that this is an application of a second-class lever!

You know that the plug is in there good and tight. But, with a mechanical advantage of ten, your 100-pound pull on the handle will twist the plug loose with a force of a half-ton.

If you've spent any time opening crates at a base, you've already used a wrecking bar. The blue-jacket in figure 1-12 is busily engaged in tearing that crate open. The wrecking bar is a first-class lever. Notice that it has curved lever arms. Can you figure the mechanical advantage of this one? Your answer should be M. A. = 5.

The crane in figure 1-13 is used for handling relatively light loads around a warehouse or a dock. You can see that the crane is rigged as a third-class lever. The effort is applied between the fulcrum and the load. This gives a mechanical advantage of less than one. If it's going to support that 1/2 ton load, you know that the pull on the lifting cable will have to be considerably greater than 1000 pounds. How much greater? Use the formula, and figure it out—

$$\frac{L}{l} = \frac{R}{E}$$

Got the answer? Right—E=1,333 lb.
Now, because the cable is pulling at an angle of about 22° at E, you can use some trigonometry to find that the pull on the cable will be about 3,560 pounds to lift the 1/2-ton weight! However, since the loads are generally light, and speed is important, it is a practical and useful machine.

Anchors are usually housed in the hawsepipe and secured by a chain stopper. The chain

7

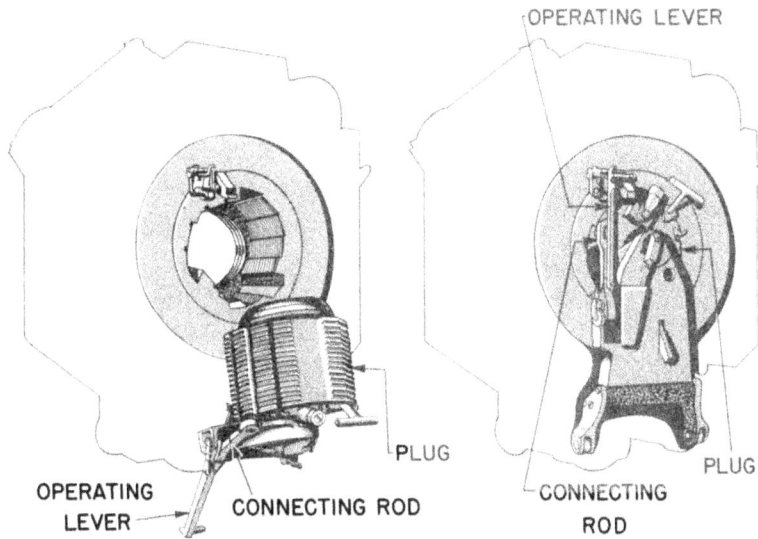

OPERATING LEVER

OPERATING
LEVER CONNECTING ROD PLUG

CONNECTING
ROD PLUG

80.213:.214

Figure 1-11.—An 8-inchers breech.

131.8

Figure 1-12.—Using a wrecking bar.

stopper consists of a short length of chain containing a turnbuckle and a pelican hook. When you secure one end of the stopper to a pad eye in the deck and lock the pelican hook over the anchor chain, the winch is relieved of the strain.

Figure 1-14A gives you the details of the pelican hook.

Figure 1-14B shows the chain stopper as a whole. Notice that the load is applied close

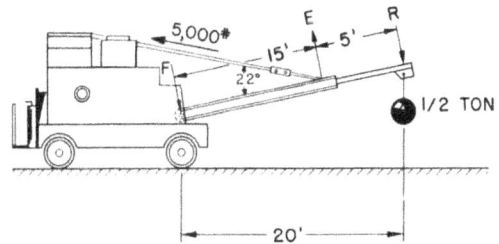

131.9

Figure 1-13.—An electric crane.

the pelican hook as a second-class lever with curved arms.

Figure 1-15 shows you a couple of guys who are using their heads to spare their muscles. Rather than exert themselves by bearing down on that drill, they pick up a board from a nearby crate and use it as a second-class lever.

If the drill is placed half way along the board, they will get a mechanical advantage of two. How would you increase the mechanical advantage if you were using this rig? Right. You move the drill in closer to the fulcrum. In the Navy, a knowledge of levers and how to apply them pays off.

3.223

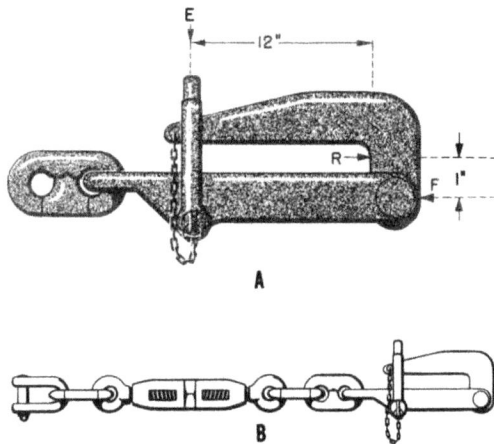

Figure 1-14.—A. A pelican hook;
B. A chain stopper.

to the fulcrum. The resistance arm a is very short. The bale shackle, which holds the hook secure, exerts its force at a considerable distance A from the fulcrum. If the chain rests against the hook one inch from the fulcrum, and the bale shackle is holding the hook closed 12 + 1 = 13 inches from the fulcrum, what's the mechanical advantage? It's 13. A strain of only 1,000 pounds on the base shackle can hold the hook closed when a 6 1/2-ton anchor is dangling over the ship's side. You'll recognize

131.10

Figure 1-15.—An improvised drill press.

CHAPTER 2

BLOCK AND TACKLE

Blocks—pulleys to a landlubber—are simple machines that have many uses aboard ship, as well as on shore. Remember how your mouth hung open as you watched movers taking a piano out of a fourth story window? The fat guy on the end of the tackle eased the piano safely to the sidewalk with a mysterious arrangement of blocks and ropes. Or perhaps you've been in the country and watched the farmer use a block-and-tackle to put hay in a barn. Since old Dobbin or the tractor did the hauling, there was no need for a fancy arrangement of ropes and blocks. Incidentally, you'll often hear the rope or tackle called the fall. Block-and-tackle, or block-and-fall.

In the Navy you'll rig a block-and-tackle to make some of your work easier. Learn the names of the parts of a block. Figure 2-1 will give you a good start on this. Look at the single block and see some of the ways you can use it. If you lash a single block to a fixed object—an overhead, a yardarm, or a bulkhead—you give yourself the advantage of being able to pull from a convenient direction. For example, in figure 2-2 you haul up a flag hoist, but you really pull down. You can do this by having a single sheaved block made fast to the yardarm. This makes it possible for you to stand in a convenient place near the flag bag and do the job. Otherwise you would have to go aloft, dragging the flag hoist behind you.

MECHANICAL ADVANTAGE

With a single fixed sheave, the force of your down-pull on the fall must be equal to the weight of the object being hoisted. You can't use this rig to lift a heavy load or resistance with a small effort—you can change only the direction of your pull.

A single fixed block is really a first-class lever with equal arms. The arms EF and FR in figure 2-3 are equal; hence the mechanical advantage is one. When you pull down at A with a force of one pound, you raise a load of one pound at B. A single fixed block does not magnify force nor speed.

You can, however, use a single block-and-fall to magnify the force you exert. Notice, in figure 2-4 that the block is not fixed, and that the fall is doubled as it supports the 200-pound cask. When rigged this way, a single block-and-fall is called a runner. Each half of the fall carries one half of the total load, or 100 pounds. Thus, by the use of the runner, the bluejacket is lifting a 200-pound cask with a 100-pound pull. The

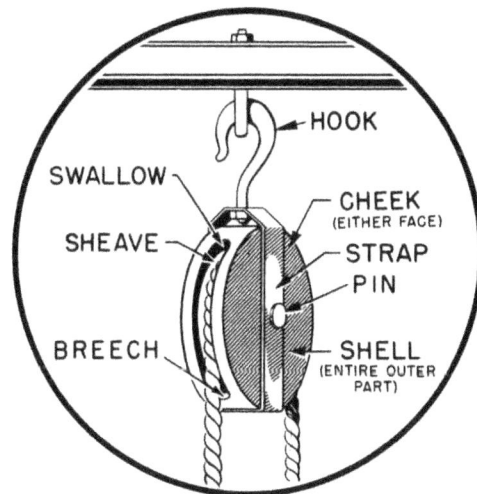

29.184

Figure 2-1.—Look it over.

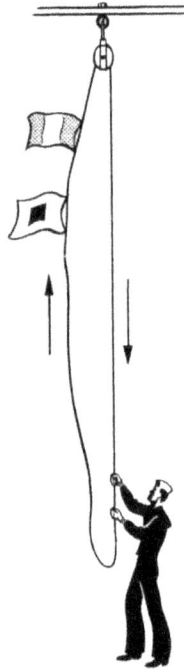

69.122

Figure 2-2.—A flag hoist.

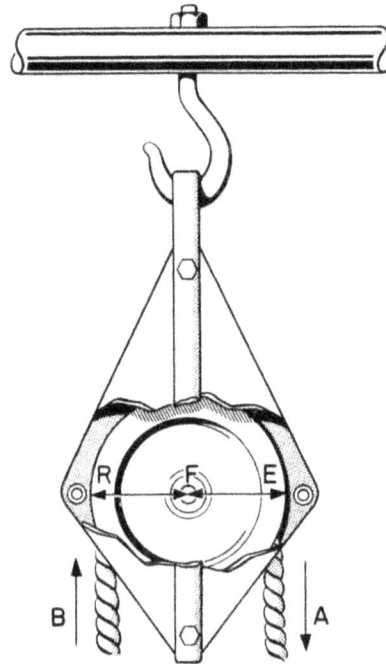

29.187

Figure 2-3.—No advantage.

mechanical advantage is two. Check this by the formula:

$$\text{M. A.} = \frac{R}{E} = \frac{200}{100}, \text{ or } 2$$

The single movable block in this setup is really a second-class lever. See figure 2-5. Your effort E acts upward upon the arm EF, which is the diameter of the sheave. The resistance R acts downward on the arm FR, which is the radius of the sheave. Since the diameter is twice the radius, the mechanical advantage is two.

But, when the effort at E moves up two feet, the load at R is raised only one foot. That's one thing to remember about blocks and falls—if you are actually getting a mechanical advantage from the system, the length of rope that passes through your hands is greater than the distance

that the load is raised. However, if you can lift a big load with a small effort, you don't care how much rope you have to pull.

The bluejacket in figure 2-4 is in an awkward position to pull. If he had another single block handy, he could use it to change the direction of the pull, as in figure 2-6. This second arrangement is known as a gun tackle purchase. Because the second block is fixed, it merely changes the direction of pull—and the mechanical advantage of the whole system remains two.

You can arrange blocks in a number of ways, depending on the job to be done and the mechanical advantage you want to get. For example, a luff tackle consists of a double block and a single block, rigged as in figure 2-7. Notice that the weight is suspended by the three parts of rope which extend from the movable single block. Each part of the rope carries its share of the load. If the crate weighs 600 pounds, then each of the three parts of the rope supports its share— 200 pounds. If there's a pull of 200 pounds downward on rope B, you will have to pull downward

11

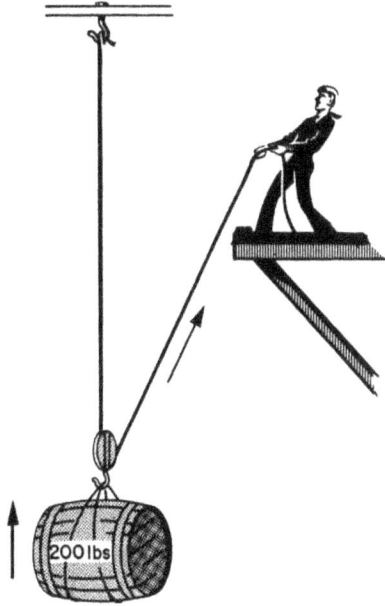

29.187

Figure 2-4.—A runner.

with a force of 200 pounds on A to counterbalance the pull on B. Neglecting the friction in the block, a pull of 200 pounds is all that is necessary to raise the crate. The mechanical advantage is:

$$M. A. = \frac{R}{E} = \frac{600}{200} = 3$$

Here's a good tip. If you count the number of the parts of rope going to and from the movable block, you can figure the mechanical advantage at a glance. This simple rule will help you to quickly approximate the mechanical advantage of most tackles you see in the Navy.

Many combinations of single, double, and triple sheave blocks are possible. Two of these combinations are shown in figure 2-8.

If you can secure the dead end of the fall to the movable block, the advantage is increased by one. Notice that this is done in figure 2-7. That is a good point to remember. Don't forget, either, that the strength of your fall—rope—is a limiting factor in any tackle. Be sure your fall will carry the load. There is no point in rigging a six-fold purchase which carries a 5-ton load with two triple blocks on a 3-inch manila rope attached to

a winch. The winch could take it, but the rope couldn't.

Now for a review of the points you have learned about blocks, and then to some practical applications aboard ship—

With a single fixed block the only advantage is the change of direction of the pull. The mechanical advantage is still one.

A single movable block gives a mechanical advantage of two.

Many combinations of single, double, and triple blocks can be rigged to give greater advantages.

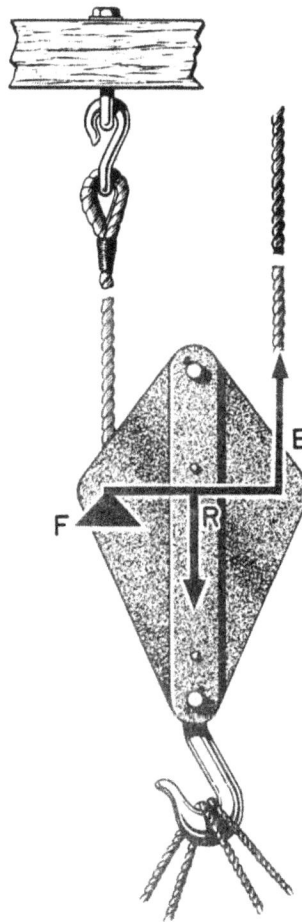

29.187

Figure 2-5.—It's 2 to 1.

12

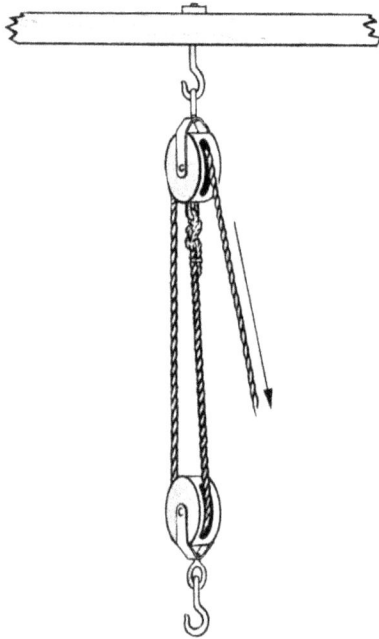

29.187

Figure 2-6.—A gun tackle.

A general rule of thumb is that the number of the parts of the fall going to and from the movable block tells you the approximate mechanical advantage of that tackle.

If you fix the dead end of the fall to the movable block you increase the mechanical advantage by one.

APPLICATIONS AFLOAT AND ASHORE

Blocks and tackle are used for a great number of lifting and moving jobs afloat and ashore. The five or six basic combinations are used over and over again in many situations. Cargo is loaded aboard, depth charges are placed in their racks, life boats are lowered over the side by the use of this machine. Heavy machinery, guns, and gun mounts are swung into position with the aid of blocks and tackle. In a thousand situations, bluejackets find this machine useful and efficient.

Yard and stay tackles are used on shipboard when you want to pick up a load from the hold and swing it onto the deck, or to shift any load a short distance. Figure 2-9 shows you how the load is first picked up by the yard tackle. The stay tackle is left slack. After the load is raised to the height necessary to clear obstructions, you take up on the stay tackle, and ease off on the yard fall. A glance at the rig tells you that the mechanical advantage of each of these tackles is only two. You may think that it isn't worth the trouble to rig a yard and stay tackle with that small advantage just to move a 400-pound crate along the deck. However, a few minutes spent in rigging may save many unpleasant hours with a sprained back.

If you want a high mechanical advantage, a luff upon luff is a good rig for you. You can raise heavy loads with this setup. Figure 2-10 shows you how it is rigged. If you apply the rule by which you count the parts of the fall going to and from the movable blocks, you find that block A gives a mechanical advantage of 3 to 1. Block B has four parts of fall running to and from it, a

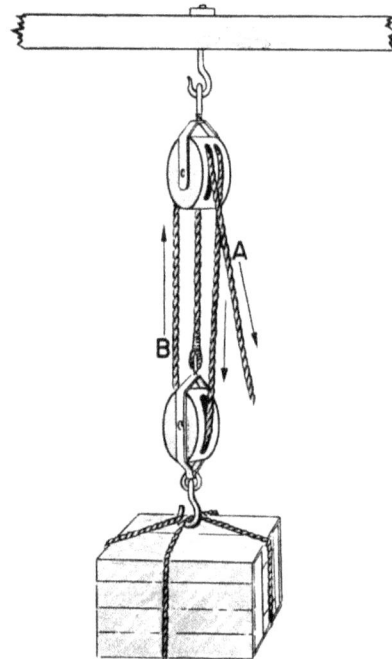

29.187

Figure 2-7.—A luff tackle.

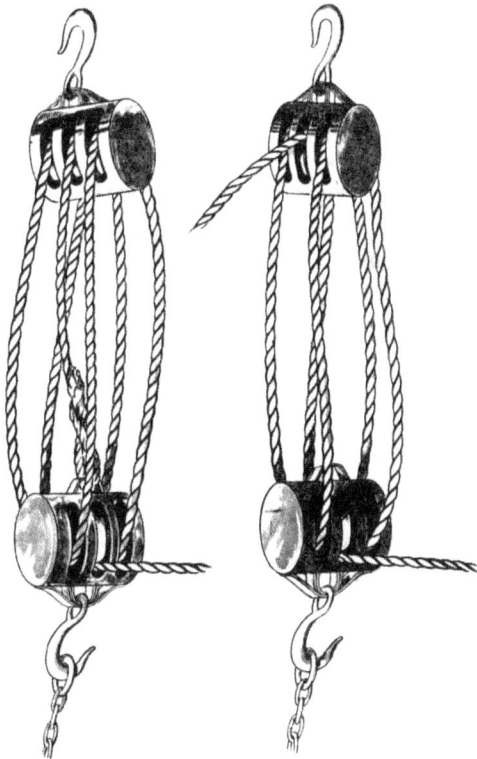

29.187

Figure 2-8.—Some other tackles.

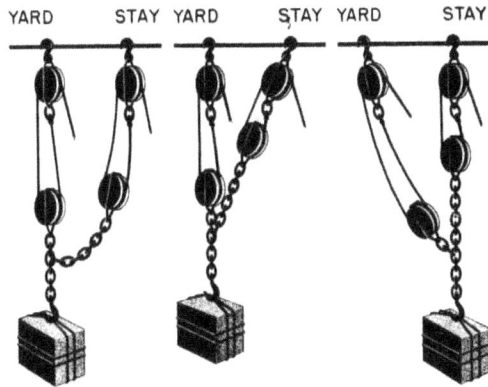

29.187

Figure 2-9.—A yard and stay tackle.

equal part of the 400-pound pull. Therefore, the hauling part requires a pull of only 1/4 x 400, or 100 pounds. So, here you have a 100-pound pull raising a 1,200-pound load. That's a mechanical advantage of 12.

In shops ashore and aboard ship you are almost certain to run into a chain hoist, or differential pulley. Ordinarily, these hoists are suspended from overhead trolleys, and are used to lift heavy objects and move them from one part of the shop to another.

mechanical advantage of 4 to 1. The mechanical advantage of those obtained from A is multiplied four times in B. The overall mechanical advantage of a luff upon luff is the product of the two mechanical advantages—or 12.

Don't make the mistake of adding mechanical advantages. Always multiply them.

You can easily figure out the M.A. for the apparatus shown in figure 2-10. Suppose the load weighs 1,200 pounds. Since it is supported by the parts 1, 2, and 3 of the fall running to and from block A, each part must be supporting one third of the load, or 400 pounds. If part 3 has a pull of 400 pounds on it, part 4 which is made fast to block B, also has a 400-pound pull on it. There are four parts of the second fall going to and from block B, and each of these takes an

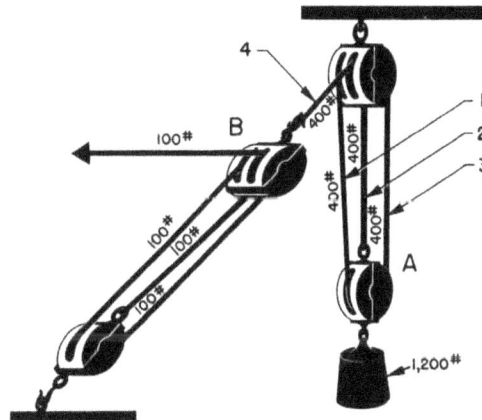

29.187

Figure 2-10.—Luff upon luff.

14

To help you to understand the operation of a chain hoist, look at the one in figure 2-11. Assume that you grasp the chain at E and pull until the large wheel A has turned around once. Then the distance through which your effort has moved is equal to the circumference of that wheel, or $2\pi R$. How much will the lower wheel C and its load be raised? Since wheel C is a single movable block, its center will be raised only one-half the distance that the chain E was pulled, or a distance πR. However, the smaller wheel B, which is rigidly fixed to A, makes one revolution at the same time as A does so B will feed some chain down to C. The length of the chain fed down will be equal to the circumference of B, or $2\pi r$. Again, since C is single movable block, the downward movement of its center will be equal to only one-half the length of the chain fed to it, or πr.

Of course, C does not first move up a distance πR and then move down a distance πr. Actually, its steady movement upward is equal to the difference between the two, or $(\pi R - \pi r)$. Don't worry about the size of the movable pulley, C. It doesn't enter into these calculations. Usually its diameter is between that of A and that of B.

The mechanical advantage equals the distance through which the effort E is moved, divided by the distance that the load is moved. This is called the velocity ratio, or theoretical mechanical advantage. It is theoretical because the frictional resistance to the movement of mechanical parts is left out. In practical uses, all moving parts have frictional resistance.

The equation for theoretical mechanical advantage may be written—

Theoretical mechanical advantage =

$$\frac{\text{Distance effort moves}}{\text{Distance resistance moves}}$$

and in this case,

$$\text{T. M. A.} = \frac{2\pi R}{\pi R - \pi r} = \frac{2R}{(R - r)}$$

If A is a large wheel, and B is a little smaller, the value of 2R becomes large, and $(R-r)$ becomes small. Then you have a large number for $\frac{2R}{(R-r)}$ which is the theoretical mechanical advantage.

You can lift heavy loads with chain hoists. To give you an idea of the mechanical advantage of a chain hoist, suppose the large wheel has a radius R of 6 inches and the smaller wheel a radius r of 5 3/4 inches. What theoretical mechanical advantage would you get? Use the formula—

$$\text{T. M. A.} = \frac{2R}{R-r}$$

Then substitute the numbers in their proper places, and solve—

$$\text{T. M. A.} = \frac{2 \times 6}{6-5\ 3/4} = \frac{12}{1/4} = 48$$

Since the friction in this type of machine is considerable, the actual mechanical advantage is not as high as the theoretical mechanical advantage would lead you to believe. For example, that theoretical mechanical advantage of 48 tells you that with a one-pound pull you should be able to lift a 48-pound load. However, actually your one-pound pull might only lift a 20-pound load. The rest of your effort would be used in overcoming the friction.

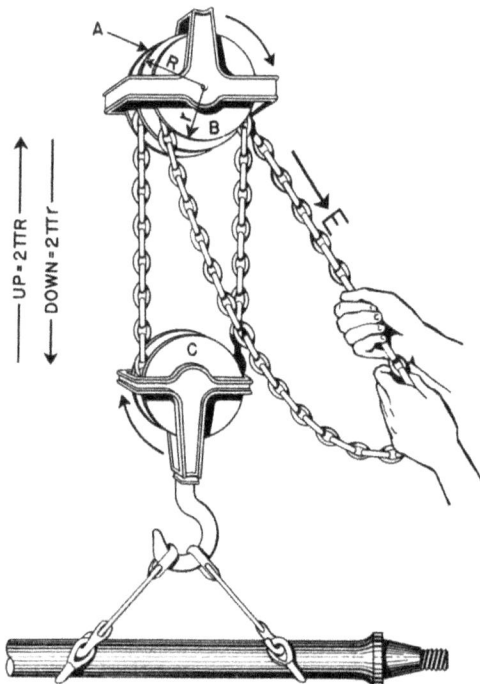

UP = 2πR — DOWN = 2πr

29.187

Figure 2-11.—A chain hoist.

CHAPTER 3

THE WHEEL AND AXLE

Have you ever tried to open a door when the knob was missing? If you have, you know that trying to twist that small four-sided shaft with your fingers is tough work. That gives you some appreciation of the advantage you get by using a knob. The door knob is an example of a simple machine called a wheel and axle.

The steering wheel on an automobile, the handle of an ice cream freezer, a brace and bit—these are familiar examples of this type of simple machine. As you know from your experience with these devices, the wheel and axle is commonly used to multiply the force you exert. If a screwdriver won't do a job because you can't turn it, you stick a screwdriver bit in the chuck of a brace and the screw probably goes in with little difficulty.

There's one thing you'll want to get straight right at the beginning. The wheel-and-axle machine consists of a wheel or crank rigidly attached to the axle, which turns with the wheel. Thus, the front wheel of an automobile is not a wheel-and-axle machine because the axle does not turn with the wheel.

MECHANICAL ADVANTAGE

How does the wheel-and-axle arrangement help to magnify the force you exert? Suppose you use a screwdriver bit in a brace to drive a stubborn screw. Look at figure 3-1A. Your effort is applied on the handle which moves in a circular path, the radius of which is 5 inches. If you apply a 10-pound force on the handle, how big a force will be exerted against the resistance at the screw? Assume the radius of the screwdriver blade is 1/4 inch. You are really using the brace as a second-class lever—see figure 3-1B. The size of the resistance

which can be overcome can be found from the formula—

$$\frac{L}{l} = \frac{R}{E}$$

In which—

L = radius of the circle through which the handle turns,
l = one-half the width of the edge of the screwdriver blade,
R = force of the resistance offered by the screw,
E = force of effort applied on the handle.

Substituting in the formula; and solving:

$$\frac{5}{1/4} = \frac{R}{10}$$

$$R = \frac{5 \times 10}{1/4}$$

$$= 5 \times 10 \times 4$$

$$= 200 \text{ lb.}$$

This means that the screwdriver blade will tend to turn the screw with a force of 200 pounds. The relationship between the radii or the diameters, or the circumferences of the wheel and axle tells you how great a mechanical advantage you can get.

Take another situation. The old oaken bucket, figure 3-2, was raised by a wheel-and-axle arrangement. If the distance from the center of the axle to the handle is 8 inches, and the radius of the drum around which the rope is wound is 2 inches, then you have a theoretical mechanical advantage of 4. That's why they used these rigs.

16

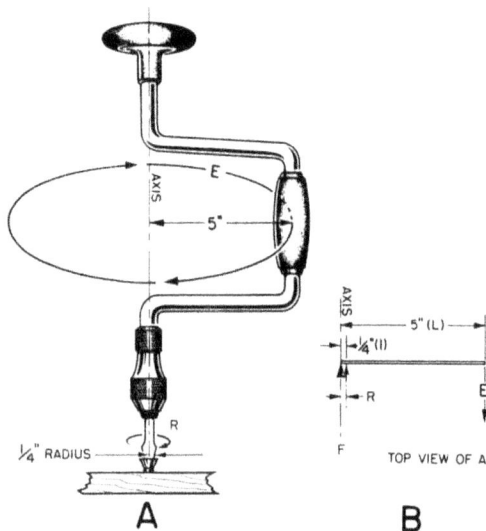

44.20

Figure 3-1.—It magnifies your effort.

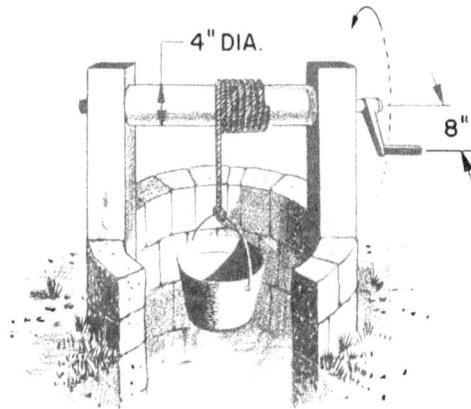

131.11

Figure 3-2.—The old oaken bucket.

MOMENT OF FORCE

In a number of situations you can use the wheel-and-axle to speed up motion. The rear-wheel sprocket of a bike, along with the rear wheel itself, is an example. When you are pedaling, the sprocket is fixed to the wheel, so the combination is a true wheel-and-axle machine. Assume that the sprocket has a circumference of 8 inches, and the wheel circumference is 80 inches. If you turn the sprocket at a rate of one revolution per second, each sprocket tooth moves at a speed of 8 inches per second. Since the wheel makes one revolution for each revolution made by the sprocket, any point on the tire must move through a distance of 80 inches in one second. So, for every eight-inch movement of a point on the sprocket, you have moved a corresponding point on the wheel through 80 inches.

Since a complete revolution of the sprocket and wheel requires only one second, the speed of a point on the circumference of the wheel is 80 inches per second, or ten times the speed of a tooth on the sprocket.

(NOTE: Both sprocket and wheel make the same number of revolutions per second so the speed of turning for the two is the same.)

Here is an idea which you will find useful in understanding the wheel and axle, as well as other machines. You probably have noticed that the force you apply to a lever tends to turn or rotate it about the fulcrum. You also know that a heave on a fall tends to rotate the sheave of the block and that turning the steering wheel of a car tends to rotate the steering column. Whenever you use a lever, or a wheel and axle, your effort on the lever arm or the rim of the wheel tends to cause a rotation about the fulcrum or the axle in one direction or another. If the rotation occurs in the same direction as the hands of a clock, that direction is called clockwise. If the rotation occurs in the opposite direction from that of the hands of a clock, the direction of rotation is called counterclockwise. A glance at figure 3-3 will make clear the meaning of these terms.

You have already seen that the result of a force acting on the handle of the carpenter's brace depends not only on the amount of that force but also on the distance from the handle to the center of rotation. From here on you'll know this result as a moment of force, or a torque (pronounced tork). Moment of force and torque have the same meaning.

Look at the effect of counterclockwise movement of the capstan bar in figure 3-4. Here the amount of the effort is designated E_1 and the distance from the point where this force is

17

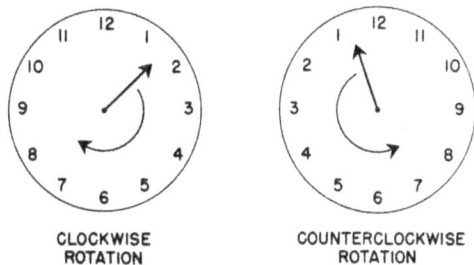

CLOCKWISE
ROTATION

COUNTERCLOCKWISE
ROTATION

131.12

Figure 3-3.—Directions of rotation.

applied to the center of the axle is L_1. Then $E_1 \times L_1$ is the moment of force. You'll notice that this term includes both the amount of the effort and the distance from the point of application of effort to the center of the axle. Ordinarily, the distance is measured in feet and the applied force is measured in pounds.

Therefore, moments of force are generally measured in foot-pounds—abbreviated ft-lb. A moment of force is frequently called a moment.

By using a longer capstan bar, the bluejacket in figure 3-4 can increase the effectiveness of his push without making a bigger effort. But if he applied his effort closer to the head of the capstan and used the same force, the moment of force would be less.

BALANCING MOMENTS

You know that the bluejacket in figure 3-4 would land flat on his face if the anchor hawser snapped. But just as long as nothing breaks, he must continue to push on the capstan bar. He is working against a clockwise moment of force, which is equal in magnitude but opposite in direction to his counterclockwise moment of force. The resisting moment, like the effort moment, depends on two factors. In the case of the resisting moment, these factors are the force R_2 with which the anchor pulls on the hawser, and the distance L_2 from the center of

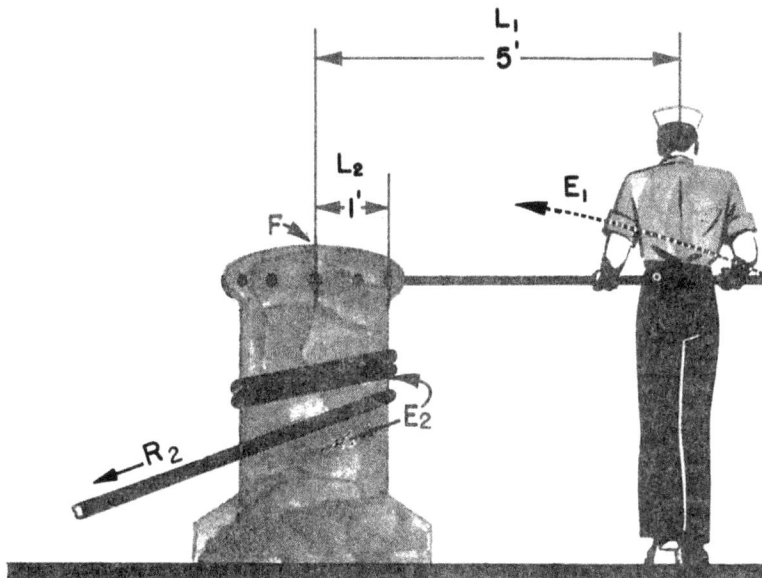

131.13

Figure 3-4.—Using a capstan.

18

the capstan to its rim. The existence of this resisting force would be evident if the blue-jacket let go of the capstan bar. The weight of the anchor pulling on the capstan would cause the whole works to spin rapidly in a clockwise direction—and good-bye anchor! The principle involved here is that whenever the counterclock-wise and the clockwise moments of force are in balance, the machine either moves at a steady speed or remains at rest.

This idea of the balance of moments of force can be summed up by the expression—

CLOCKWISE MOMENTS =
COUNTERCLOCKWISE MOMENTS

And, since a moment of force is the product of the amount of the force times the distance the force acts from the center of rotation, this expression of equality may be written—

$$E_1 \times L_1 = E_2 \times L_2$$

In which—
E_1 = force of effort,
L_1 = distance from fulcrum or axle to point where force is applied,
E_2 = force of resistance,
L_2 = distance from fulcrum or center of axle to the point where resistance is applied.

EXAMPLE 1

Put this formula to work on a capstan prob-lem. A single capstan bar is gripped 5 feet from the center of a capstan head with a radius of one foot. A 1/2-ton anchor is to be lifted. How big a push does the sailor have to exert?

First, write down the formula—

$$E_1 \times L_1 = E_2 \times L_2$$

Here L_1 = 5; E_2 = 1,000 pounds; and L_2 = 1. Substitute these values in the formula, and it becomes:

$$E_1 \times 5 = 1,000 \times 1$$

and—

$$E_1 = \frac{1,000}{5} = 200 \text{ pounds}$$

EXAMPLE 2

Consider now the sad case of Slim and Sam, as illustrated in figure 3-5. Slim has suggested that they carry the 300-pound crate slung on a

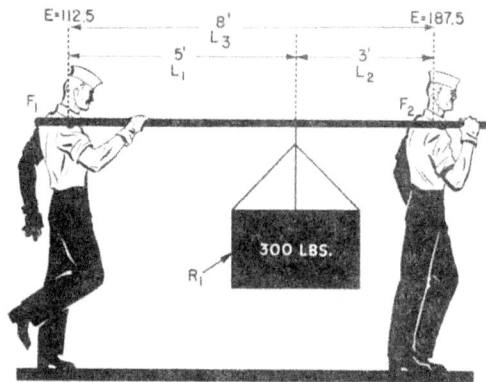

131.14
Figure 3-5.—A practical application.

handy 10-foot pole. He was smart enough to slide the load up 3 feet from Sam's shoulder.

Here's how they made out. Use Slim's shoulder as a fulcrum F_1. Look at the clock-wise moment caused by the 300-pound load. That load is five feet away from Slim's shoulder. If R_1 is the load, and L_1 the distance from Slim's shoulder to the load, the clockwise moment M_A is—

$$M_A = R_1 \times L_1 = 300 \times 5 = 1,500 \text{ ft-lb.}$$

With Slim's shoulder still acting as the ful-crum, the resistance of Sam's effort causes a counterclockwise moment M_B acting against the load moment. This counterclockwise moment is equal to Sam's effort E_2 times the distance L_3 from his shoulder to the fulcrum F_1 at Slim's shoulder. Since L_2 = 8 ft., the formula is—

$$M_B = E_2 \times L_3 = E_2 \times 8 = 8E_2$$

But there is no rotation, so the clockwise moment and the counterclockwise moment are equal. $M_A = M_B$. Hence—

$$1,500 = 8E_2$$

$$E_2 = \frac{1,500}{8} = 187.5 \text{ pounds.}$$

So poor Sam is carrying 187.5 pounds of the 300-pound load.

19

What is Slim carrying? The difference between 300 and 187.5 = 112.5 pounds, of course! But you can check your answer by this procedure—

This time, use Sam's shoulder as the fulcrum F_2. The counterclockwise moment M_c is equal to the 300-pound load R_1 times the distance L_2—3 feet—from Sam's shoulder, or M_c 300 x 3 = 900 foot-pounds. The clockwise moment M_D is the result of Slim's lift E_1 acting at a distance L_3 from the fulcrum. L_3 = 8 feet. Again, since counterclockwise moment equals clockwise moment, you have—

$$900 = E_1 \times 8$$

and
$$E_1 = \frac{900}{8} = 112.5 \text{ pounds}$$

Slim, the smart sailor, has to lift only 112.5 pounds. There's a blue jacket who really puts his knowledge to work.

THE COUPLE

Take a look at figure 3-6. It's another capstan-turning situation. To increase effective effort, a second capstan bar is placed opposite the first and another bluejacket can apply a force on the second bar. The two sailors in figure 3-6 will apparently be pushing in opposite directions. But, since they are on opposite sides of the axle, they are actually causing rotation in the same direction. And, if the two sailors are pushing with equal forces, the moment of force is twice as great as though only one sailor was pushing. This arrangement is known technically as a couple.

You will see that the couple is a special example of the wheel and axle. The moment of force is equal to the product of the total distance L_T between the two points of effort and the force E_1 applied by one sailor. The equation for the couple may be written

$$E_1 \times L_T = E_2 \times L_2$$

SUMMARY

Here is a quick review of the wheel and axle— things you should have straight in your mind—

A wheel-and-axle machine has the wheel fixed rigidly to the axle. The wheel and the axle turn together.

The wheel and axle may be used either to magnify your effort or to speed it up.

The effect of a force tending to rotate an object around an axis or fulcrum is called a moment of force, or simply a moment.

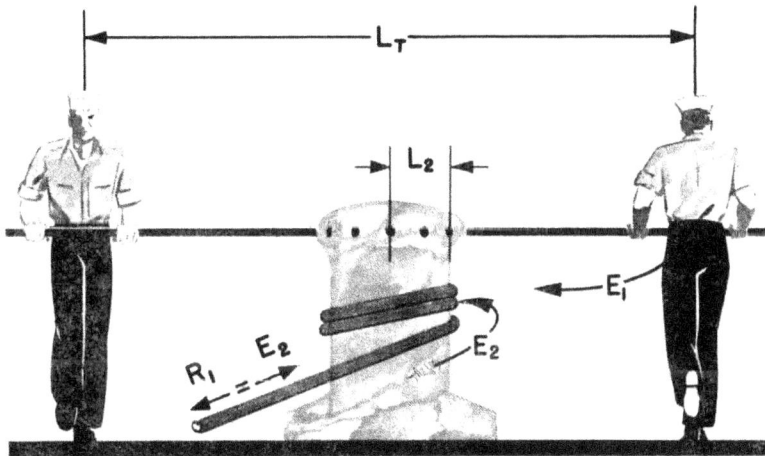

131.15

Figure 3-6.—A couple.

When an object is at rest or is moving steadily, the clockwise moments are just equal and opposite to the counterclockwise moments.

Moments of the force, depend upon two factors—the amount of the force, and the distance from the fulcrum or axis to the point where the force is applied.

When two equal forces are applied at equal distances on opposite sides of a fulcrum, and move in opposite directions so that they both tend to cause rotation about the fulcrum, you have a couple.

APPLICATIONS AFLOAT AND ASHORE

A trip to the engine room makes you realize how important the wheel and axle is on the modern ship. Everywhere you look you see wheels of all sizes and shapes. Most of them are used to open and close valves quickly. One common type of valve is shown in figure 3-7. Turning the wheel causes the threaded stem to rise and open the valve. Since the valve must close water-tight, air-tight, or steam-tight, all the parts must fit snugly. To move the stem on most valves without the aid of the wheel would be impossible. The wheel gives you the necessary mechanical advantage.

You've handled enough wrenches to know that the longer the handle, the tighter you can turn a nut. Actually, a wrench is a wheel-and-axle machine. You can consider the handle as one spoke of a wheel, and the place where you take hold of the handle as a point on the rim. The nut which is held in the jaws of the wrench can be compared to the axle.

You know that you can turn a nut too tight and strip the threads or cause internal parts to seize. This is especially true when you are taking up on bearings. In order to make the proper adjustment, you use a torque wrench. There are several types. Figure 3-8 shows you one that is very simple. When you pull on the handle, its shaft bends. The rod on which the pointer is fixed does not bend—so the pointer indicates on the scale the torque, or moment of force, that you are exerting. The scale is generally stated in pounds, although it is really measuring foot-pounds of torque. If the nut is to be tightened by a moment of 90 ft-lb, you pull until the pointer is opposite the number 90 on the scale. The servicing or repair manual on an engine or piece of machinery generally tells you what the torque—or moment of force—should be on each set of nuts or bolts.

The gun pointer uses a couple to elevate and depress the gun barrel. He cranks away at a hand-wheel that has two handles. The right-hand handle is on the opposite side of the axle from the left-hand handle—180° apart. Look at figure 3-9. When he pulls on one handle and pushes on the other, he's producing a couple. But if he lets go the left handle to scratch himself, and cranks only with his right hand, he no longer has a couple—just a simple first-class lever! And he'd have to push twice as hard with one hand.

131.16

Figure 3-7.—Valves.

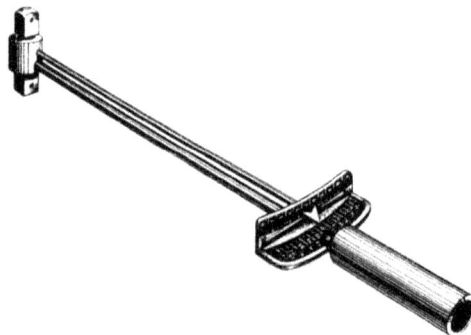

5.9

Figure 3-8.—A simple torque wrench.

A system of gears—a gear train—transmits the motion to the barrel. A look at figure 3-10 will help you to figure the forces involved. The radius of the wheel is 6 inches—1/2 foot—and each handle is being turned with a force of, say, 20 pounds. The moment on the top which tends to rotate the wheel in a clockwise direction is equal to 20 times 1/2 = 10 ft-lb. The bottom handle also rotates the wheel in the same direction with an equal moment. Thus the total twist or torque on the wheel is 10 + 10 = 20 ft-lb. To get the same moment with one hand, applying a 20-pound force, the radius of the wheel would have to be twice as great—12 inches, or one foot. The couple is a convenient arrangement of the wheel-and-axle machine.

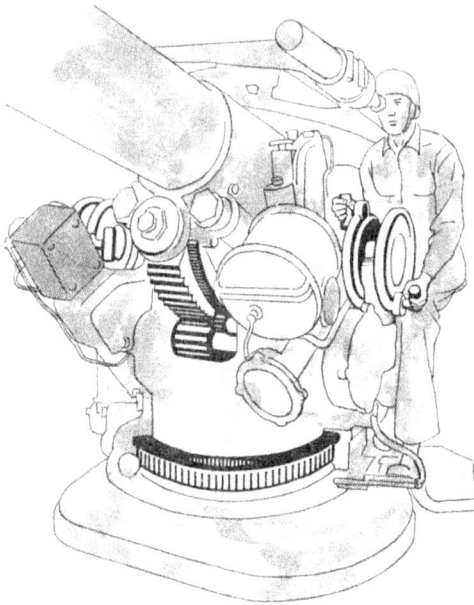

131.17

Figure 3-9.—A pointer's handwheel.

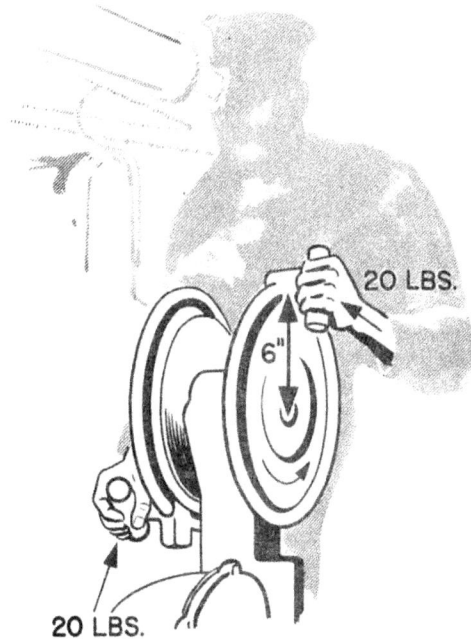

131.18

Figure 3-10.—Developing a torque.

CHAPTER 4

THE INCLINED PLANE AND THE WEDGE

THE BARREL ROLL

You have probably watched a driver load barrels on a truck. The truck is backed up to the curb. The driver places a long double plank or ramp from the sidewalk to the tail gate, and then rolls the barrel up the ramp. A 32-gallon barrel may weigh close to 300 pounds when full, and it would be quite a job to lift one up into the truck. Actually, the driver is using a simple machine called the inclined plane. You have seen the inclined plane used in many situations. Cattle ramps, a mountain highway, and the gangplank are familiar examples.

The inclined plane permits you to overcome a large resistance by applying a relatively small force through a longer distance than the load is raised. Look at figure 4-1. Here you see the driver easing the 300-pound barrel up to the bed of the truck, three feet above the sidewalk. He is using a plank nine feet long. If he didn't use the ramp at all, he'd have to apply a 300-pound force straight up through the three-foot distance. With the ramp, however, he can apply his effort over the entire nine feet of the plank as the barrel is slowly rolled up to a height of three feet. It looks, then, as if he could use a force only three-ninths of 300, or 100 pounds, to do the job. And that is actually the situation.

Here's the formula. Remember it from chapter 1?

$$\frac{L}{l} = \frac{R}{E}$$

In which— L = length of the ramp, measured along the slope,
l = height of the ramp,
R = weight of object to be raised, or lowered,
E = force required to raise or lower object

Now apply the formula to this problem—

In this case, L = 9 ft.; l = 3 ft; and R = 300 lb. By substituting these values in the formula, you get—

$$\frac{9}{3} = \frac{300}{E}$$

$$9E = 900$$

$$E = 100 \text{ pounds}$$

Since the ramp is three times as long as its height, the mechanical advantage is three. You find the theoretical mechanical advantage by dividing the total distance through which your effort is exerted by the vertical distance through which the load is raised or lowered.

THE WEDGE

The wedge is a special application of the inclined plane. You have probably used wedges. Abe Lincoln used a wedge to help him split logs into rails for fences. The blades of knives, axes, hatchets, and chisels act as wedges when they are forced into a piece of wood. The wedge is two inclined planes, set base-to-base. By driving the wedge full-length into the material to be cut or split, the material is forced apart a distance equal to the width of the broad end of the wedge. See figure 4-2.

Long, slim wedges give high mechanical advantage. For example, the wedge of figure 4-2 has a mechanical advantage of six. Their greatest value, however, lies in the fact that you can use them in situations where other simple machines won't work. Imagine the trouble you'd have trying to pull a log apart with a system of pulleys.

23

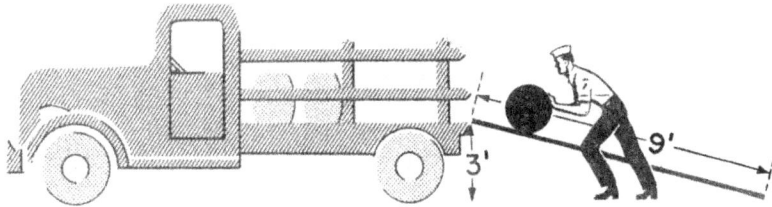

110.4

Figure 4-1.—An inclined plane.

SUMMARY

Before you look at some of the Navy applications of the inclined plane and the wedge, here's a summary of what to remember from this chapter—

The inclined plane is a simple machine that lets you raise or lower heavy objects by applying a small force over a relatively long distance.

The theoretical mechanical advantage of the inclined plane is found by dividing the length of the ramp by the perpendicular height that the load will be raised or lowered. The actual mechanical advantage is equal to the weight of the resistance or load, divided by the force that must be used to move the load up the ramp.

The wedge is two inclined planes set base-to-base. It finds its greatest use in cutting or splitting materials.

APPLICATIONS AFLOAT AND ASHORE

One of the most common uses of the inclined plane in the Navy is the gangplank. Going aboard the ship by gangplank, illustrated in figure 4-3 is certainly easier than climbing up a sea ladder. And you appreciate the M.A. of the gangplank even more when you have to carry your sea bag or a case of prunes aboard.

Remember that hatch dog in figure 1-10. The dog that's used to secure a door not only takes advantage of the lever principle, but—if you look sharply—you can see that the dog seats itself on a steel wedge which is welded to the door. As the dog slides upward along this wedge, it forces the door tightly shut. This is an inclined plane, with its length about eight times its thickness. That means you get a theoretical mechanical advantage of eight. You figured, in

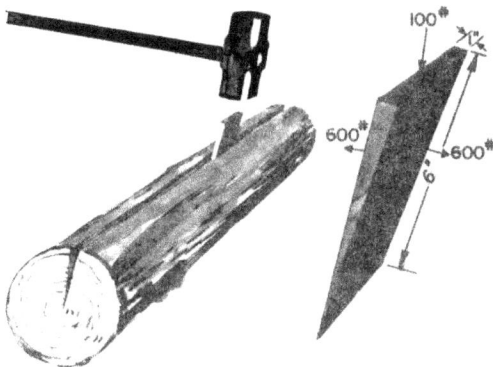

131.19

Figure 4-2.—A wedge.

131.20

Figure 4-3.—The gangplank is an inclined plane.

chapter 1, that you got a mechanical advantage of four from the lever action of the dog—so the overall mechanical advantage is 8 times 4 or 32, neglecting friction. Not bad for such a simple gadget, is it? Push down with 50 pounds heave on the handle and you squeeze the door shut with a force of 1600 pounds, on that dog. You'll find the damage-control parties using wedges by the dozen to shore up bulkheads and decks. A few sledge-hammer blows on a wedge will quickly and firmly tighten up the shoring.

Chipping scale or paint off steel is a tough job. However, the job is made a lot easier with a compressed air chisel. The wedge-shaped cutting edge of the chisel gets in under the scale or the paint, and exerts great pressure to lift the scale or paintlayer. The chisel bit is another application of the inclined plane.

CHAPTER 5

THE SCREW

A MODIFIED INCLINED PLANE

The screw is a simple machine that has many uses. The vise on a workbench makes use of the great mechanical advantage of the screw. So do the screw clamps used to hold a piece of furniture together while it is being glued. And so do many automobile jacks and even the food grinder in the kitchen at home.

A screw is a modification of the inclined plane. Cut a sheet of paper in the shape of a right triangle—an inclined plane. Wind it around a pencil, as in figure 5-1. Then you can see that the screw is actually an inclined plane wrapped around a cylinder. As the pencil is turned, the paper is wound up so that its hypotenuse forms a spiral thread similar to the thread on the screw shown at the right. The pitch of the screw, and of the paper, is the distance between identical points on the same threads, and measured along the length of the screw.

THE JACK

In order to understand how the screw works, look at figure 5-2. Here you see a jack screw of the type that is used to raise a house or a piece of heavy machinery. The jack has a lever handle with a length R. If you pull the lever handle around one turn, its outer end has described a circle. The circumference of this circle is equal to 2π. (You remember that π equals 3.14, or $\frac{22}{7}$). That is the distance, or the lever arm, through which your effort is applied.

At the same time, the screw has made one revolution, and in doing so has been raised a height equal to its pitch P. You might say that one full thread has come up out of the base. At any rate, the load has been raised a distance P.

Remember that the theoretical mechanical advantage is equal to the distance through which the effort or pull is applied, divided by the distance the resistance or load is moved. Assuming a 2-foot—24"—length for the lever arm, and a 1/4-inch pitch for the thread, you can find the theoretical mechanical advantage by the formula—

$$\text{M. A.} = \frac{2\pi r}{p}$$

in which

R = length of handle = 24 inches
P = pitch, or distance between corresponding points on successive threads = 1/4-inch.

Substituting,

$$\text{T. M. A.} = \frac{2 \times 3.14 \times 24}{1/4} = \frac{150.72}{1/4} = 602.88$$

A 50-pound pull on the handle would result in a theoretical lift of 50x602 or about 30,000 pounds. Fifteen tons for fifty pounds.

But jacks have considerable friction loss. The threads are cut so that the force used to overcome friction is greater than the force used to do useful work. If the threads were not cut this way, if no friction were present, the weight of the load would cause the jack to spin right back down to the bottom as soon as you released the handle.

THE MICROMETER

In using the jack, you exerted your effort through a distance of $2\pi r$, or 150 inches, in order to raise the screw 1/4 inch. It takes a lot of circular motion to get a small amount of straight-line motion from the head of the jack. You will use this point to advantage in the micrometer, which is a useful device for making accurate small measurements, measurements of a few thousandths of an inch.

In figure 5-3, you see a cutaway view of a micrometer. The thimble turns freely on the

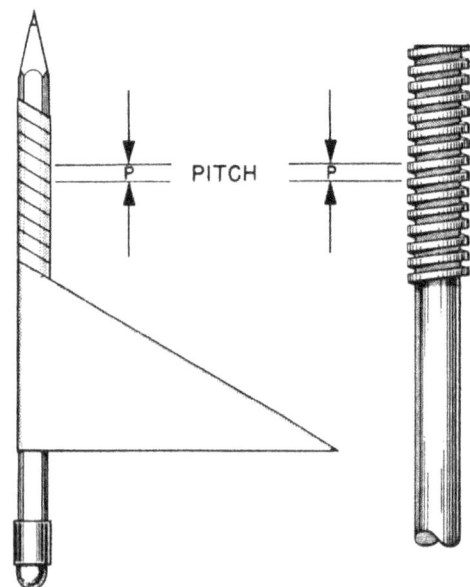

110.4

Figure 5-1.—A screw is an inclined plane in spiral form.

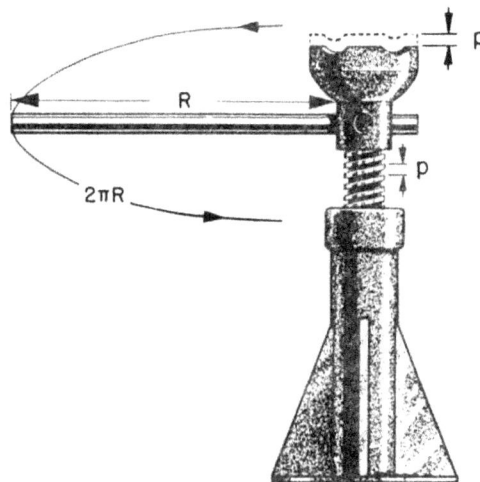

81.25

Figure 5-2.—A jack screw.

sleeve, which is rigidly attached to the micrometer frame. The spindle is attached to the thimble, and is fitted with screw threads which move the spindle and thimble to right or left in the sleeve when you rotate the thimble. These screw threads are cut 40 threads to the inch. Hence one turn of the thimble moves the spindle and thimble 1/40 inch. This represents one of the smallest divisions on the micrometer. Four of these small divisions make 4/40 of an inch, or 1/10 inch. Thus the distance from 0 to 1 or 1 to 2 on the sleeve represents 1/10 or 0.1 inch.

To allow even finer measurements, the thimble is divided into 25 equal parts laid out by graduation marks around its rim, as shown in figure 5-4. If you turn the thimble through 25 of these equal parts, you have made one complete revolution of the screw, which represents a lengthwise movement of 1/40 of an inch. Now, if you turn the thimble one of these units on its scale, you have moved the spindle a distance of 1/25 of 1/40 inch, or 1/1000 of an inch—0.001 inch.

The micrometer in figure 5-4 reads 0.503 inch, which is the true diameter of the half-inch drill-bit shank being measured. This tells you that the diameter of this particular bit is 0.003 inch greater than its nominal diameter of 1/2 inch—0.500''.

Because you can make such accurate measurements with this instrument, it is indispensable in every machine shop.

SUMMARY

Look over the basic ideas you have learned from this chapter, and then see how the Navy uses this simple machine—the screw.

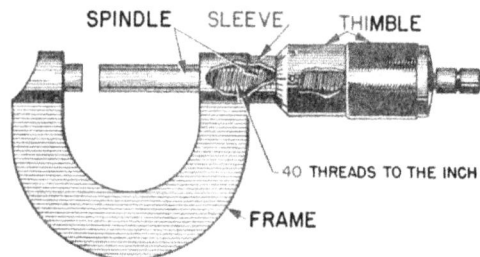

4.20

Figure 5-3.—A micrometer.

4.20

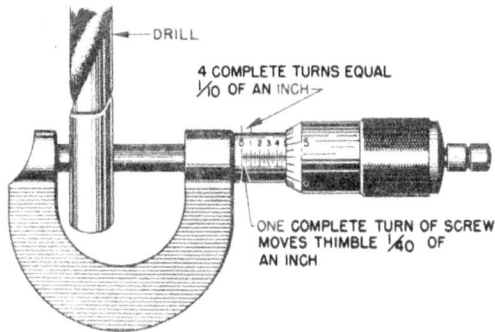

Figure 5-4.—Taking turns.

The screw is a modification of the inclined plane—modified to give you a high mechanical advantage.

The theoretical mechanical advantage of the screw can be found by the formula

$$M.A. = \frac{2\pi r}{P}$$

As in all machines, the actual mechanical advantage equals the resistance divided by the effort.

In many applications of the screw, you make use of the large amount of friction that is commonly present in this simple machine.

By the use of the screw, large amounts of circular motion are reduced to very small amounts of straight-line motion.

APPLICATIONS AFLOAT AND ASHORE

It's a tough job to pull a rope or cable up tight enough to get all the slack out of it. But you can do it. Use a turnbuckle. The turnbuckle is an application of the screw. See figure 5-5. If you turn it in one direction, it takes up the slack in a cable. Turning it the other way slacks off on the cable. You'll notice that one bolt of the turnbuckle has left-hand threads, and the other bolt has right-hand threads. Thus, when you turn the turnbuckle to tighten up the line, both bolts tighten up. If both bolts were right-hand thread—standard thread—one would tighten while the other one loosened an equal amount. Result—no change in cable-slack. Most turnbuckles have the screw threads cut to provide a large amount of frictional resistance to keep the turnbuckle from unwinding under load. In some cases, the turnbuckle has a lock nut on each of the screws to prevent slipping. You'll find turnbuckles used in a hundred different ways afloat and ashore.

Ever wrestled with a length of wire rope? Obstinate and unwieldly, wasn't it? Riggers have dreamed up tools to help subdue wire rope. One of these tools—the rigger's vise—is shown in figure 5-6. This rigger's vise uses the great mechanical advantage of the screw to hold the wire rope while the crew splices a thimble—a reinforced loop—onto the end of the cable. Rotating the handle causes the jaw on that screw to move in or out along its grooves. This machine is a modification of the vise on a work bench. Notice the right-hand and left-hand screws on the left-hand clamp.

Figure 5-7 shows you another use of the screw. Suppose you want to stop a winch with its load suspended in mid-air. To do this, you need a brake. The brake on most anchor or cargo winches consists of a metal band that

131.21

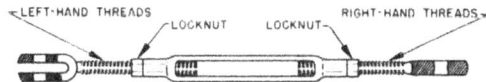

Figure 5-5.—A turnbuckle.

131.22

Figure 5-6.—A rigger's vise.

28

131.23

Figure 5-7.—A friction brake.

encircles the brake drum. The two ends of the band are fastened to nuts connected by a screw attached to a handwheel. As you turn the handwheel, the screw pulls the lower end of the band A up toward its upper end B. The huge M.A. of the screw puts the squeeze on the drum, and all rotation of the drum is stopped.

One type of steering gear used on many small ships—and as a spare steering system on some larger ships—is the screw gear. Figure 5-8 shows you that the wheel turns a long threaded shaft. Half the threads—those nearer the wheel end of this shaft—are right-hand threads. The other half of the threads—those farther from the wheel—are left-hand threads. The nut A has a right-hand thread, and nut B has a left-hand thread. Notice that the cross head which turns the rudder is connected to the nuts by two steering arms. If you stand in front of the wheel and turn it in a clockwise direction—to your right—arm A moves forward and arm B moves backward. This turns the rudder counterclockwise, so that the ship swings in the direction you turn the wheel. There is a great mechanical advantage to this steering mechanism.

Figure 5-9 shows you another practical use of the screw. The quadrant davit makes it possible for two men to put a large life boat over the side with little effort. The operating handle is attached to a threaded screw which passes through a traveling nut. If the operating handle is cranked in a counterclockwise direction (as you face outboard), the nut travels outward along the screw. The traveling nut is fastened to the davit arm by a swivel. The davit arm and the boat swing outboard as a result of the outward movement of the screw. The thread on that screw is the self-locking type—if you let go of the handle the nut remains locked in position.

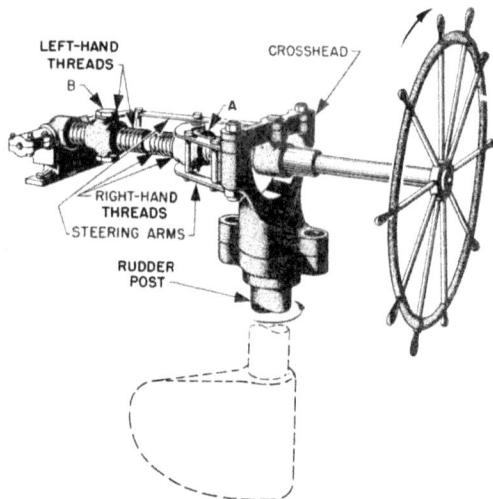

131.24

Figure 5-8.—The screw gives a tremendous mechanical advantage.

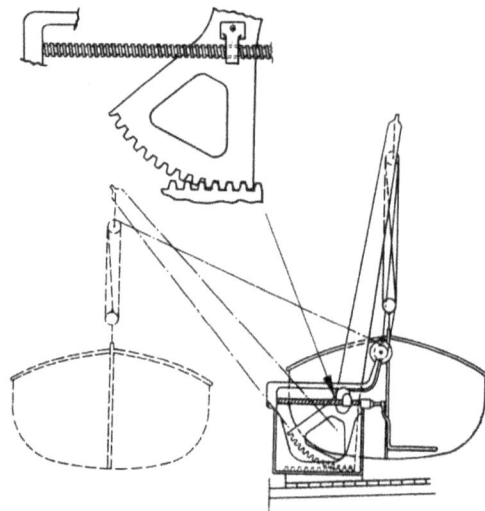

80.101

Figure 5-9.—The quadrant davit.

CHAPTER 6

GEARS

Did you ever take a clock apart to see what made it tick? Of course you came out with some parts left over when you got it back together again. And they probably included a few gear wheels. Gears are used in many machines. Frequently the gears are hidden from view in a protective case filled with grease or oil, and you may not see them.

An egg beater gives you a simple demonstration of the three things that gears do. They can change the direction of motion; increase or decrease the speed of the applied motion; and magnify or reduce the force which you apply. Gears also give you a positive drive. There can be, and usually is, creep or slip in a belt drive. But gear teeth are always in mesh, and there can be no creep and slip.

Follow the directional changes in figure 6-1. The crank handle is turned in the direction indicated by the arrow—clockwise, when viewed from the right. The 32 teeth on the large vertical wheel A mesh with the 8 teeth on the right-hand horizontal wheel B, which rotates as indicated by the arrow. Notice that as B turns in a clockwise direction, its teeth mesh with those of wheel C and cause wheel C to revolve in the opposite direction. The rotation of the crank handle has been transmitted by gears to the beater blades, which also rotate.

Now figure out how the gears change the speed of motion. There are 32 teeth on gear A and 8 teeth on gear B. But the gears mesh, so that one complete revolution of A results in four complete revolutions of gear B. And since gears B and C have the same number of teeth, one revolution of B results in one revolution of C. Thus the blades revolve four times as fast as the crank handle.

In chapter 1 you learned that third-class levers increase speed at the expense of force. The same thing happens with this egg beater. The magnitude of the force is changed. The force required to turn the handle is greater than the force applied to the frosting by the blades. Therefore a mechanical advantage of less than one results.

TYPES OF GEARS

When two shafts are not lying in the same straight line, but are parallel, motion can be transmitted from one to the other by means of spur gears. This setup is shown in figure 6-2.

Spur gears are wheels with mating teeth cut in their surfaces so that one can turn the other without slippage. When the mating teeth are cut so that they are parallel to the axis of rotation, as shown in figure 6-2, the gears are called straight spur gears.

When two gears of unequal size are meshed together, the smaller of the two is usually called a pinion. By unequal size, we mean an unequal number of teeth causing one gear to be of a larger diameter than the other. The teeth, themselves, must be of the same size in order to mesh properly.

The most commonly used type are the straight spur gears, but quite often you'll run across another type of spur gear called the helical spur gear.

In helical gears the teeth are cut slantwise across the working face of the gear. One end of the tooth, therefore, lies ahead of the other. In other words, each tooth has a leading end and a trailing end. A look at these gears in figure 6-3A will show you how they're constructed.

In the straight spur gears the whole width of the teeth comes in contact at the same time. But with helical (spiral) gears contact between two teeth starts first at the leading ends and moves progressively across the gear faces until the trailing ends are in contact. This kind of meshing action keeps the gears in constant contact with one another. Therefore, less lost

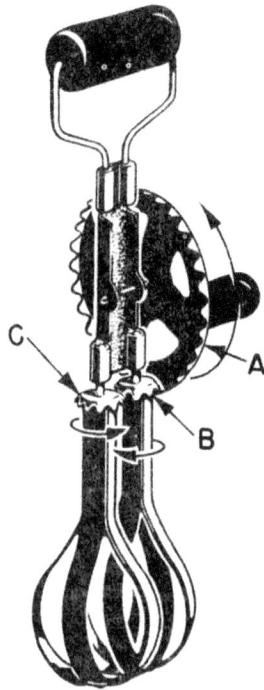

131.25
Figure 6-1.—A simple gear arrangement.

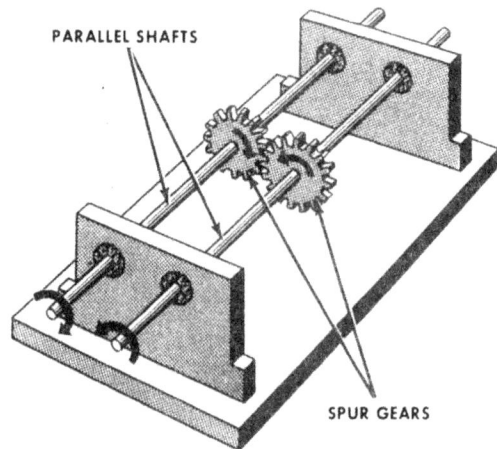

5.22.1
Figure 6-2.—Spur gears coupling two parallel shafts.

motion and smoother, quieter action is possible. One disadvantage of this helical spur gear is the tendency of each gear to thrust or push axially on its shaft. It is necessary to put a special thrust bearing at the end of the shaft to counteract this thrust.

Thrust bearings are not needed if herringbone gears like those shown in figure 6-4 are used. Since the teeth on each half of the gear are cut in opposite directions, each half of the gear develops a thrust which counterbalances that of the other half. You'll find herringbone gears used mostly on heavy machinery.

Figure 6-3 also shows you three other gear arrangements in common use.

The internal gear in figure 6-3B has teeth on the inside of a ring, pointing inward toward the axis of rotation. An internal gear is always meshed with an external gear, or pinion, whose

center is offset from the center of the internal gear. Either the internal or pinion gear can be the driver gear, and the gear ratio is calculated the same as for other gears—by counting teeth.

Often only a portion of a gear is needed where the motion of the pinion is limited. In this case the sector gear (fig. 6-3C) is used to save space and material. The rack and pinion in figure 6-3D are both spur gears. The rack may be considered as a piece cut from a gear with an extremely large radius. The rack-and-pinion arrangement is useful in changing rotary motion into linear motion.

THE BEVEL GEAR.—So far most of the gears you've learned about transmit motion between parallel shafts. But when shafts are not parallel (at an angle), another type of gear is used—the bevel gear. This type of gear can connect shafts lying at any given angle because they can be beveled to suit the angle.

Figure 6-5A shows a special case of the bevel gear—the miter gear. A pair of miter gears is used to connect shafts having a 90° angle, which means the gear faces are beveled at a 45° angle.

You can see in figure 6-5B how bevel gears are designed to join shafts at any angle. Gears cut at any angle other than 45° are called just plain bevel gears.

31

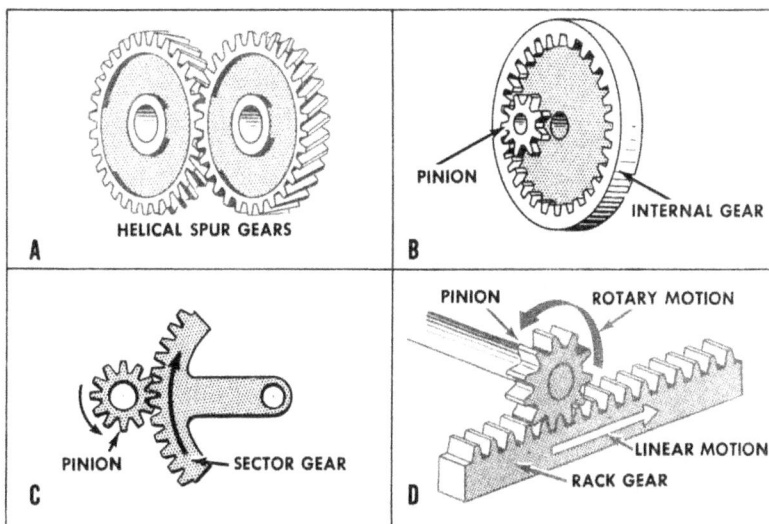

5.23:.24

Figure 6-3.—Gear types.

The gears shown in figure 6-5 are called straight bevel gears, because the whole width of each tooth comes in contact with the mating tooth at the same time. However, you'll also run across spiral bevel gears with teeth cut so as to have advanced and trailing ends. Figure 6-6 shows you what spiral bevel gears look like. They have the same advantages as other spiral (helical) gears—less lost motion and smoother, quieter operation.

THE WORM AND WORM WHEEL.—Worm and worm-wheel combinations, like those in figure 6-7, have many uses and advantages. But it's better to understand their operating theory before learning of their uses and advantages.

Figure 6-7A shows the action of a single-thread worm. For each revolution of the worm,

5.22.3

Figure 6-4.—Herringbone gear.

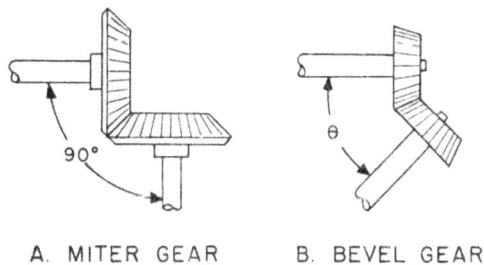

A. MITER GEAR   B. BEVEL GEAR

5.22.4

Figure 6-5.—Bevel gears.

32

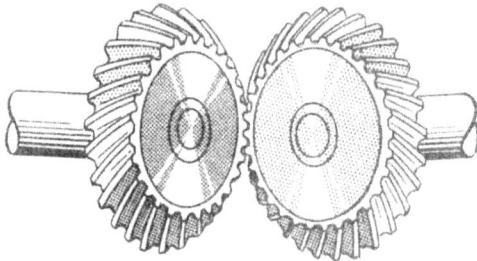

5.22.6

Figure 6-6.—Spiral bevel gears.

the worm wheel turns one tooth. Thus if the worm wheel has 25 teeth the gear ratio is 25:1.

Figure 6-7B shows a double-thread worm. For each revolution of the worm in this case, the worm wheel turns two teeth. That makes the gear ratio 25:2 if the worm wheel has 25 teeth.

Likewise, a triple-threaded worm would turn the worm wheel three teeth per revolution of the worm.

A worm gear is really a combination of a screw and a spur gear. Tremendous mechanical advantages can be obtained with this arrangement. Worm drives can also be designed so that only the worm is the driver—the

spur cannot drive the worm. On a hoist, for example, you can raise or lower the load by pulling on the chain which turns the worm. But if you let go of the chain, the load cannot drive the spur gear and let the load drop to the deck. This is a non-reversing worm drive.

CHANGING DIRECTION WITH GEARS

No doubt you know that the crankshaft in an automobile engine can turn in only one direction. If you want the car to go backwards, the effect of the engine's rotation must be reversed. This is done by a reversing gear in the transmission, not by reversing the direction in which the crankshaft turns.

A study of figure 6-8 will show you how gears are used to change the direction of motion. This is a schematic diagram of the sight mounts on a Navy gun. If you crank the range-adjusting handle A in a clockwise direction the gear B directly above it is made to rotate in a counterclockwise direction. This motion causes the two pinions C and D on the shaft to turn in the same direction as gear B against the teeth cut in the bottom of the table. The table is tipped in the direction indicated by the arrow.

As you turn the deflection-adjusting handle E in a clockwise direction the gear F directly above it turns in the opposite direction. Since the two bevel gears G and H are fixed on the

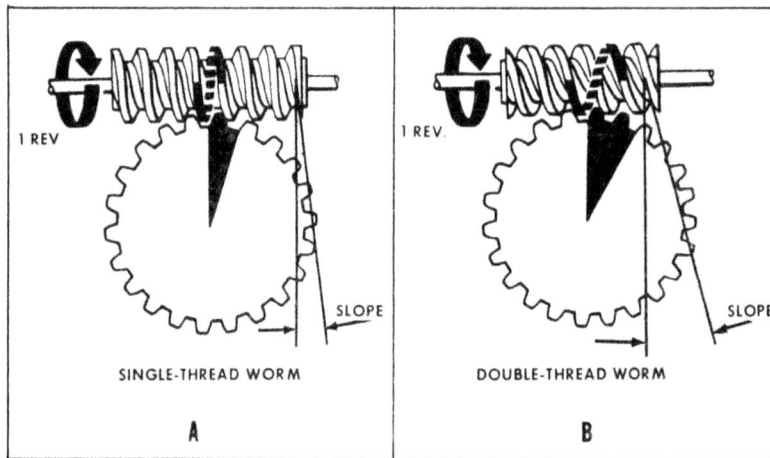

5.22.9

Figure 6-7.—Worm gears.

33

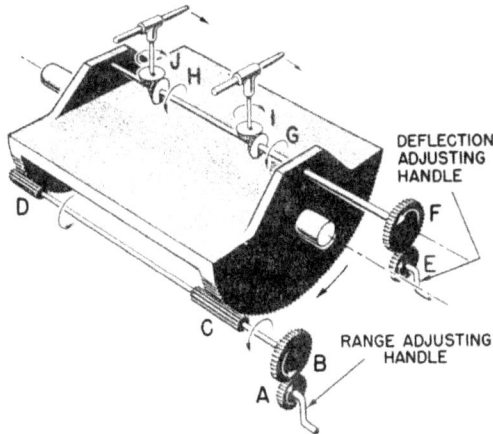

131.26

Figure 6-8.—Gears change direction of
applied motion.

shaft with F, they also turn. These bevel gears,
meshing with the horizontal bevel gears I and
J, cause I and J to swing the front ends of the
telescopes to the right. Thus with a simple
system of gears, it is possible to keep the two
telescopes pointed at a moving target. In this
and many other practical applications, gears
serve one purpose—to change the direction of
motion.

CHANGING SPEED

As you've already seen in the egg-beater,
gears can be used to change the speed of motion.
Another example of this use of gears is found
in your clock or watch. The mainspring slowly
unwinds and causes the hour hand to make one
revolution in 12 hours. Through a series—or
train—of gears, the minute hand makes one
revolution each hour, while the second hand
goes around once per minute.

Figure 6-9 will help you to understand how
speed changes are made possible. Wheel A
has 10 teeth which mesh with the 40 teeth on
wheel B. Wheel A will have to rotate four
times to cause B to make one revolution.
Wheel C is rigidly fixed on the same shaft with
B. Thus C makes the same number of revolutions as B. However, C has 20 teeth, and
meshes with wheel D which has only 10 teeth.
Hence, wheel D turns twice as fast as wheel C.

Now, if you turn A at a speed of four revolutions per second, B will be rotated at one
revolution per second. Wheel C also moves
at one revolution per second, and causes D to
turn at two revolutions per second. You get
out two revolutions per second after having
put in four revolutions per second. Thus the
overall speed reduction is 2/4—or 1/2—which
means that you got half the speed out of the
last driven wheel that you put into the first
driver wheel.

You can solve any gear speed-reduction
problem with this formula—

$$S_2 = S_1 \times \frac{T_1}{T_2},$$

where

S_1 = speed of first shaft in train
S_2 = speed of last shaft in train
T_1 = product of teeth on all drivers
T_2 = product of teeth on all driven gears

Now use the formula on the gear train of
figure 6-9.

$$S_2 = S_1 \times \frac{T_1}{T_2} = 4 \times \frac{10 \times 20}{40 \times 10} =$$

$$\frac{800}{400} = 2 \text{ revs. per sec.}$$

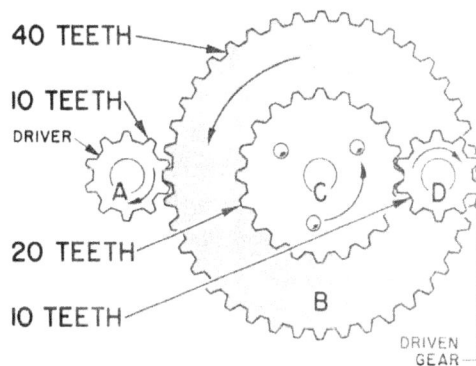

131.27

Figure 6-9.—Gears can change speed of
applied motion.

Almost any increase or decrease in speed can be obtained by choosing the correct gears for the job. For example, the turbines on a ship have to turn at high speeds—say 5800 rpm—if they are going to be efficient. But the propellers, or screws, must turn rather slowly—say 195 rpm—to push the ship ahead with maximum efficiency. So, a set of reduction gears is placed between the turbines and the propeller shaft.

When two external gears mesh, they rotate in opposite directions. Often you'll want to avoid this. Put a third gear, called an idler, between the driver and the driven gear. But don't let this extra gear confuse you on speeds. Just neglect the idler entirely. It doesn't change the gear ratio at all, and the formula still applies. The idler merely makes the driver and its driven gear turn in the same direction. Figure 6-10 will show you how this works.

MAGNIFYING FORCE WITH GEARS

Gear trains are used to increase the mechanical advantage. In fact, wherever there is a speed reduction, the effect of the effort you apply is multiplied. Look at the cable winch in figure 6-11. The crank arm is 30 inches long, and the drum on which the cable is wound has a 15-inch radius. The small pinion gear has 10

teeth, which mesh with the 60 teeth on the internal spur gear. You will find it easier to figure the mechanical advantage of this machine if you think of it as two machines.

First, figure out what the gear and pinion do for you. The theoretical mechanical advantage of any arrangement of two meshed gears can be found by the following formula—

$$M. A. (theoretical) = \frac{T_o}{T_a}$$

In which, T_o = number of teeth on driven gear;
T_a = number of teeth on driver gear.
In this case, T_o = 60 and T_a = 10. Then,

$$M. A. (theoretical) = \frac{T_o}{T_a} = \frac{60}{10} = 6$$

Now, for the other part of the machine, which is a simple wheel-and-axle arrangement consisting of the crank arm and the drum. The theoretical mechanical advantage of this can be found by dividing the distance the effort moves—$2\pi R$—in making one complete revolution, by the distance the cable is drawn up in one revolution of the drum—$2\pi r$.

$$M. A. (theoretical) = \frac{2\pi R}{2\pi r} = \frac{R}{r} = \frac{30}{15} = 2$$

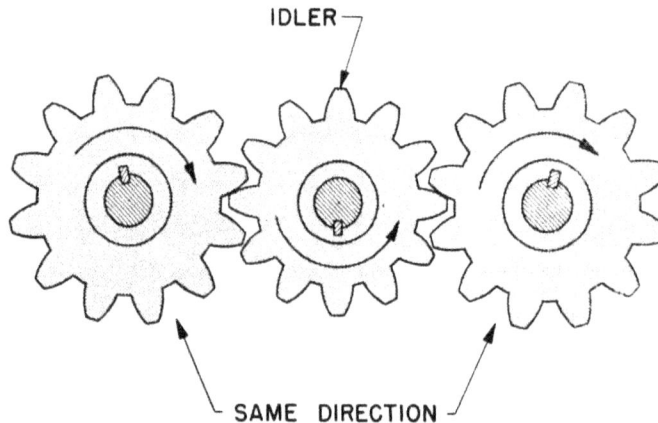

Figure 6-10.—An idler gear.

12.55

35

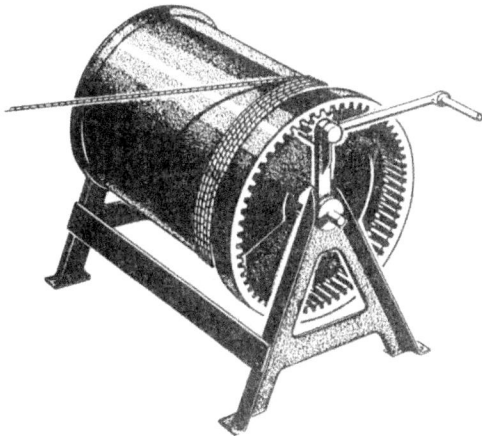

131.28

Figure 6-11.—This magnifies your effort.

You know that the total, or overall, theoretical mechanical advantage of a compound machine is equal to the product of the mechanical advantages of the several simple machines that make it up. In this case you considered the winch as being two machines—one having an M. A. of 6, and the other an M. A. of 2. Therefore, the over-all theoretical mechanical advantage of the winch is 6 x 2, or 12. Since friction is always present, the actual mechanical advantage may be only 7 or 8. Even so, by applying a force of 100 pounds on the handle, you could lift a load of 700 or 800 pounds.

You use gears to produce circular motion. But you often want to change rotary motion into up-and-down or linear motion. You can use cams to do this. For example—

The cam shaft in figure 6-12 is turned by the gear. A cam is keyed to the shaft and turns with it. The cam has an irregular shape which is designed to move the valve stem up and down, giving the valve a straight-line motion as the cam shaft rotates.

When the cam shaft rotates, the high point—lobe—of the cam raises the valve to its open position. As the shaft continues to rotate, the high point of the cam is passed and the valve is lowered to closed position.

A set of cams, two to a cylinder, driven by timing gears from the crankshaft operate the exhaust and intake valves on the gasoline automobile engine as shown in figure 6-13. Cams are widely used in machine tools and other devices to make rotating gears and shafts do up-and-down work.

SUMMARY

These are the important points you should keep in mind about gears—

Gears can do a job for you by changing the direction, speed, or size of the force which you apply.

When two external gears mesh, they always turn in opposite directions. You can make them turn in the same direction by placing an idler gear between the two.

The product of the number of teeth on each of the driver gears, divided by the product of the number of teeth on each of the driven gears, gives you the speed ratio of any gear train.

The theoretical mechanical advantage of any gear train is the product of the number of teeth on the driven gear wheels, divided by the product of the number of teeth on the driver gears.

The overall theoretical mechanical advantage of a compound machine is equal to the product of the theoretical mechanical advantages of all the simple machines which make it up.

Cams are used to change rotary motion into linear motion.

One of the gear systems you'll get to see frequently aboard ship is that on the anchor winch. Figure 6-14 shows you one type in which you can readily see how the wheels go 'round. The driving gear A is turned by the winch engine or motor. This gear has 22 teeth, which mesh with the 88 teeth on the large wheel B. Thus, you know that the large wheel makes one revolution for every four revolutions of the driving gear A. You get a 4-to-1 theoretical mechanical advantage out of that pair. Secured to the same shaft with B is the small spur gear C, covered up here. The gear C has 30 teeth which mesh with the 90 teeth on the large gear D, also covered up. The advantage from C to D is

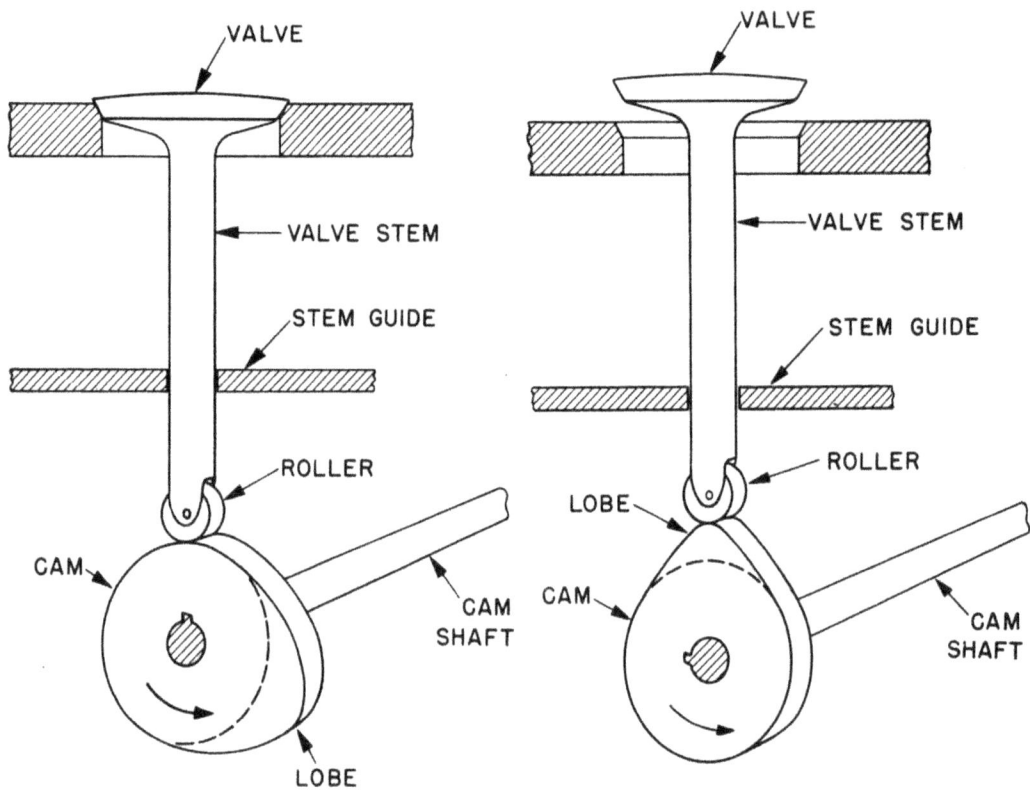

131.29

Figure 6-12.—Cam-driven valve.

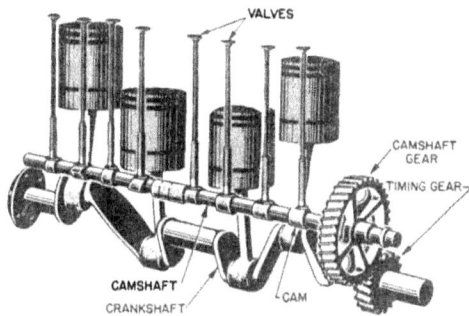

131.30

Figure 6-13.—Automobile valve gear.

3 to 1. The sprocket wheel to the far left, on the same shaft with D, is called a wildcat. The anchor chain is drawn up over this. Every second link is caught and held by the protruding teeth of the wildcat. The overall mechanical advantage of the winch is 4 x 3, or 12 to 1.

Figure 6-15 shows you an application of the rack and pinion as a steering mechanism. Turning the ship's wheel turns the small pinion A. This pinion causes the internal spur gear to turn. Notice that there is a large mechanical advantage in the arrangement.

Now you see that center pinion P turns. It meshes with the two vertical racks. When the wheel is turned full to the right, one rack moves downward and the other moves upward to the positions of the racks. Attached to the bottom of the racks are two hydraulic pistons which control the steering of the ship. You'll get some information on this hydraulic system in a later chapter.

131.31

Figure 6-14.—An anchor winch.

131.32

Figure 6-15.—A steering mechanism.

CHAPTER 7

WORK

MEASUREMENT

You know that machines help you to do work. But just what is work? Work doesn't mean simply applying a force. If that were so, you would have to consider that Big-Boy, busily applying his 220-pound force on the sea bag in figure 7-1 is doing work. But no work is being done!

Work, in the mechanical sense of the term, is done when a resistance is overcome by a force acting through a measureable distance. Now, if Big-Boy were to lift his 90-pound bag off the deck and put it on his bunk, he would be doing work. He would be overcoming a resistance by applying a force through a distance.

Notice that two factors are involved—force and movement through a distance. The force is normally measured in pounds, and the distance in feet. Work, therefore, is commonly measured in units called foot-pounds. You do one foot-pound of work when you lift a one-pound weight through a height of one foot. But—you also do one foot-pound of work when you apply one pound of force on any object through a distance of one foot. Writing this as a formula, it becomes—

$$\frac{\text{WORK}}{\text{(foot-pounds)}} = \frac{\text{FORCE}}{\text{(pounds)}} \times \frac{\text{DISTANCE}}{\text{(feet)}}$$

Thus, if the sailor lifts a 90-pound bag through a vertical distance of 5 feet, he will do—

$$\text{WORK} = 90 \times 5 = 450 \text{ ft -lb.}$$

There are two points concerning work that you should get straight right at the beginning.

First, in calculating the work done you measure the actual resistance being overcome. This is not necessarily the weight of the object being moved. To make this clear, look at the job the bluejacket in figure 7-2 is doing. He is pulling a 900-pound load of supplies 200 feet along the dock. Does this mean that he is doing 900 times 200, or 180,000 foot-pounds of work? Of course not. He isn't working against the pull of gravity—or the total weight—of the load. He's pulling only against the rolling friction of the truck, and that may be as little as 90 pounds. That is the resistance which is being overcome. Always be sure that you know what resistance is being overcome by the effort, as well as the distance through which it is moved. The resistance in one case may be the weight of the object; in another it may be the frictional resistance of the object as it is dragged or rolled along the deck.

The second point to hold in mind is that you have to move the resistance to do any work on it. Look at Willie in figure 7-3. The poor guy has been holding that suitcase for the past 15 minutes waiting for the bus. His arm is getting tired; but according to the definition of work, he isn't doing any—because he isn't moving the suitcase. He is merely exerting a force against the pull of gravity on the bag.

You already know about the mechanical advantage of a lever. Now consider it in terms of getting work done easily. Look at figure 7-4. The load weighs 300 pounds, and you want to lift it up onto a platform a foot above the deck. How much work must you do on it? Since 300 pounds must be raised one foot, 300 times 1, or 300 foot-pounds of work must be done. You can't make this weight any smaller by the use of any machine. However, if you use the eight-foot plank as shown, you can do that amount of work, by applying a smaller force through a longer distance. Notice that you have a mechanical advantage of 3, so that a 100-pound push down on the end of the plank will raise the 300-pound crate. Through how long a distance will you have to exert that 100-pound push? Neglecting

NO LOITERING

131.33

Figure 7-1.—No work is being done.

friction—and in this case you can safely do so—the work done on the machine is equal to the work done by the machine. Say it this way—

Work put in = work put out.

And since Work = force x distance, you can substitute "force times distance" on each side of the work equation. Thus—

$$F_1 \text{ times } S_1 = F_2 \text{ times } S_2$$

in which,

F_1 = effort applied, in pounds
S_1 = distance through which effort moves, in feet
F_2 = resistance overcome, in pounds
S_2 = distance resistance is moved, in feet

Now substitute the known values, and you obtain—

$$100 \text{ times } S_1 = 300 \text{ times } 1$$

$$S_1 = 3 \text{ feet}$$

131.34

Figure 7-2.—Working against friction.

131.35

Figure 7-3.—No motion no work.

The advantage of using the lever is not that it makes any less work for you, but that it allows you to do the job with the force at your command. You'd probably have some difficulty lifting 300 pounds directly upward without a machine to help you!

A block and tackle also makes work easier. But like any other machine, it can't decrease the total amount of work to be done. With a rig like the one shown in figure 7-5, the bluejacket has a mechanical advantage of 5, neglecting friction. Notice that five parts of the rope go to and from the movable block. To raise the 600-pound load 20 feet, he needs to exert a pull of only 1/5 of 600—or 120 pounds. But—he is going to have to pull more than 20 feet of rope through his hands in order to do this. Use the formula again to figure why this is so—

Work input = work output

$$F_1 \times S_1 = F_2 \times S_2$$

And by substituting the known values—

$$120 \times S = 600 \times 20$$

$$S_1 = 100 \text{ feet.}$$

This means that he has to pull 100 feet of rope through his hands in order to raise the load 20 feet. Again, the advantage lies in the fact that a relatively small force operating through a large distance can move a big load through a small distance.

The sailor busy with the big piece of machinery in figure 7-6 has his work cut out for him. He is trying to seat the machine square on its foundations. The rear end must be shoved over one-half foot against a frictional resistance of 1,500 pounds. The amount of work to be done is 1,500x1/2, or 750 foot-pounds. He will have to do at least this much work on the jack he is using. If the jack has a 2 1/2-foot handle—R = 2 1/2 ft — and the pitch of the jack screw is 1/4 inch, he can do the job with little effort. Neglecting friction, you can figure it out this way—

Work input = work output

$$F_1 \times S_1 = F_2 \times S_2$$

In which

F_1 = force in pounds applied on the handle;
S_1 = distance, in feet, that the end of the handle travels in one revolution;
F_2 = resistance to be overcome;
S_2 = distance in feet that head of jack is advanced by one revolution of the screw. Or, the pitch of the screw.

And, by substitution,

$$F_1 \times 2R = 1500 \times 1/48,$$

since 1/4" = 1/48 of a foot

$$F_1 \times 2 \times 2 1/2 = 1500 \times 1/48$$

$$F_1 = 2 \text{ pounds}$$

The jack makes it theoretically possible for the sailor to exert a 1,500-pound push with a 2-pound effort, but look at the distance through which he must apply that effort. One complete turn of the handle represents a distance of 15.7 feet. That 15.7-foot rotation advances the piece of machinery only 1/4th of an inch—or 1/48th of a foot. Force is gained at the expense of distance.

41

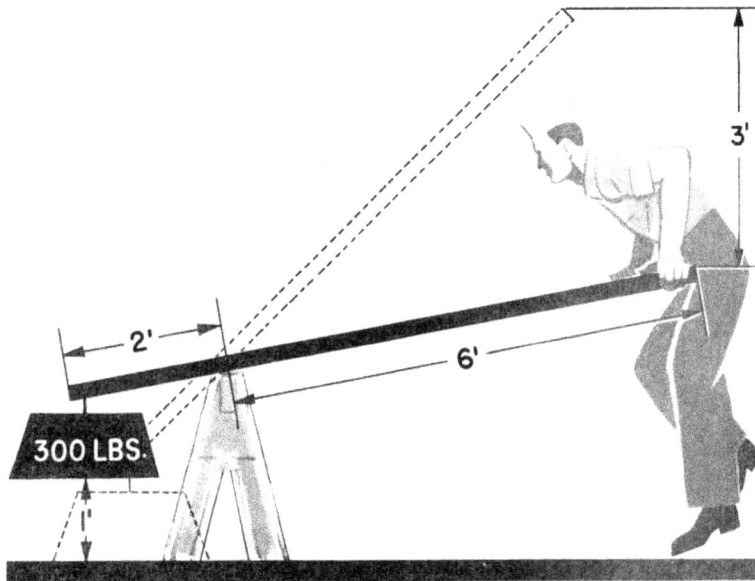

131.36

Figure 7-4.—Push'em up.

FRICTION

You are going to push a 400-pound crate up a 12-foot plank, the upper end of which is 3 feet higher than the lower end. You figure out that a 100-pound push will do the job, since the height the crate is to be raised is one-fourth of the distance through which you are exerting your push. The theoretical mechanical advantage is 4. And then you push 100 pounds worth. Nothing happens! You've forgotten that there is friction between the surface of the crate and the surface of the plank. This friction acts as a resistance to the movement of the crate—and you must overcome this resistance to move the crate. In fact, you might have to push as much as 150 pounds to move it. Fifty pounds would be used to overcome the frictional resistance, and the remaining 100 pounds would be the useful push that would move the crate up the plank.

Friction is the resistance that one surface offers to its movement over another surface. The amount of friction depends upon the nature of the two surfaces, and the forces which hold them together.

In many instances friction is useful to you. Friction helps you hold back the crate from sliding down the inclined ramp. The cinders you throw under the wheels of your car when it's slipping on an icy pavement increase the friction. You wear rubber-soled shoes in the gym to keep from slipping. And locomotives carry a supply of sand which can be dropped on the tracks in front of the driving wheels to increase the friction between the wheels and the track. Nails hold structures together because of the friction between the nails and the lumber.

When you are trying to stop or slow up an object in motion, when you want traction, and when you want to prevent motion from taking place, you make friction work for you. But when you want a machine to run smoothly and at high efficiency, you eliminate as much friction as possible by oiling and greasing bearings, and honing and smoothing rubbing surfaces.

Wherever you apply force to cause motion, friction makes the actual mechanical advantage fall short of the theoretical mechanical advantage. Because of friction, you have to make a greater effort to overcome the resistance which you want to move. If you place a marble and a

131.37

Figure 7-5.—A block and tackle makes
work easier.

131.38

Figure 7-6.—A big push.

lump of sugar on a table and give each an equal push, the marble will move farther. This is because rolling friction is always less than sliding friction. You take advantage of this fact whenever you use ball bearings or roller bearings. See figure 7-7.

Remember that rolling friction is always less than sliding friction. The Navy takes advantage of that fact. Look at figure 7-8. This roller chock not only cuts down the wear and tear on lines and cables which are run through it, but—by reducing friction—also reduces the load the winch has to work against.

The roller bitt in figure 7-9 is another example of how you can cut down the wear and tear on lines or cable and also reduce your frictional loss.

When it is necessary to have one surface move over another, you can decrease the friction by the use of lubricants, such as oil, grease, or soap. You will use lubricants on flat surfaces, gun slides for example, as well as on ball and roller bearings, to further reduce the frictional resistance and to cut down the wear.

Don't forget that in a lot of situations friction is mighty helpful, however. Many a blue-jacket has found out about this the hard way—on a wet, slippery deck. On some of our ships you'll find that a rough-grained deck covering is used. Here you have friction working for you. It helps you to keep your footing.

EFFICIENCY

Up to this point you have been neglecting the effect of friction on machines. This makes it easier to explain machine operation, but you know from practical experience that friction is involved every time two surfaces move against one another. And the work used in overcoming the frictional resistance does not appear in the work output. Since this is so, it's obvious that you have to put more work into a machine than you get out of it. In other words, no machine is 100 percent efficient.

Take the jack in figure 7-6, for example. The chances are good that a 2-pound force exerted on the handle wouldn't do the job at all. More likely a pull of at least 10 pounds would be required. This indicates that only 2 out of the 10 pounds, or 20 percent of the effort is usefully employed to do the job. The remaining 8 pounds of effort were consumed in overcoming the friction in the jack. Thus, the jack has an efficiency of only 20 percent. Most jacks are inefficient, but even with this inefficiency, it is possible to deliver a huge push with a small amount of effort.

43

Actually, $50 \div 0.60 = 83.3$ pounds. You can check this yourself in the following manner—

$$\text{Efficiency} = \frac{\text{Output}}{\text{Input}}$$

$$= \frac{F_2 \times S_2}{F_1 \times S_1}$$

One revolution of the drum would raise the 600-pound load a distance S_2 of $2\pi r$ or 7.85 feet. To make the drum revolve once, the pinion gear must be rotated six times by the handle; and the handle must be turned through a distance S_1 of $6 \times 2\pi R$, or 94.2 feet. Then, by substitution—

$$0.60 = \frac{600 \times 7.85}{F_1 \times 94.2}$$

$$F_1 = \frac{600 \times 7.85}{94.2 \times 0.60} = 83.3 \text{ pounds.}$$

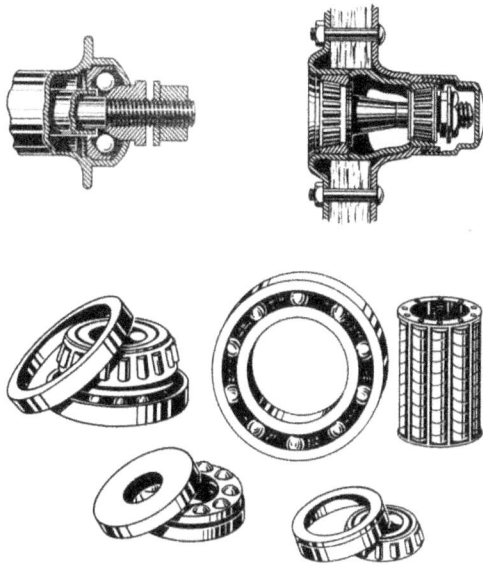

12.50

Figure 7-7.—These reduce friction.

A simple way to calculate the efficiency of a machine is to divide the output by the input—convert to percentage

$$\text{Efficiency} = \frac{\text{Output}}{\text{Input}}$$

Now go back to the block-and-tackle problem illustrated in figure 7-5. It's likely that instead of being able to lift the load with a 120-pound pull, the bluejacket would perhaps have to use a 160-pound pull through the 100 feet. You can calculate the efficiency of the rig by the following method—

$$\text{Efficiency} = \frac{\text{output}}{\text{input}} \qquad\qquad = \frac{F_2 \times S_2}{F_1 \times S_1}$$

and, by substitution,

$$= \frac{600 \times 20}{160 \times 100} = 0.75 \text{ or } 75 \text{ percent.}$$

Theoretically, with the mechanical advantage of twelve developed by the cable winch back in figure 6-11 you should be able to lift a 600-pound load with a 50-pound push on the handle. If the machine has an efficiency of 60 percent, how big a push would you actually have to apply?

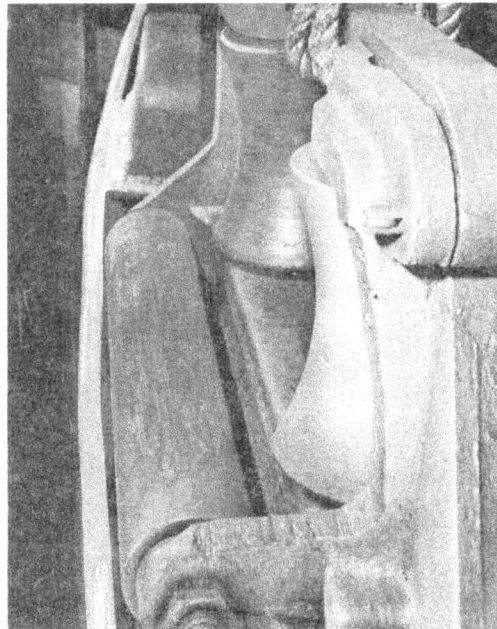

131.39

Figure 7-8.—It saves wear and tear.

Because this machine is only 60-percent efficient, you have to put 94.2x83.3, or 7,847 footpounds of work into it in order to get 4,710 foot-pounds of work out of it. The difference—7,847—4,710=3,137 foot-pounds—is used to overcome friction within the machine.

SUMMARY

Here are some of the important points you should remember about friction, work, and efficiency—

You do work when you apply a force against a resistance and move the resistance.

Since force—measured in pounds—and distance—measured in feet—are involved, work is measured in foot-pounds. One foot-pound of work is the result of a one-pound force, acting against a resistance through a distance of one foot.

Machines help you to do work by making it possible to move a large resistance through a small distance by the application of a small force through a large distance.

Since friction is present in all machines, more work must be done on the machine than the machine actually does on the load.

The efficiency of any machine can be found by dividing the output by the input.

The resistance that one surface offers to movement over a second surface is called friction.

Friction between two surfaces depends upon the nature of the materials and the magnitude of the forces pushing them together.

131.40

Figure 7-9.—Roller bitt saves line.

45

CHAPTER 8

POWER

It's all very well to talk about how much work a man can do, but the payoff is how long it takes him to do it. Look at "Lightning" in figure 8-1. He has lugged 3 tons of bricks up to the second deck of the new barracks. However, it has taken him three 10-hour days—1800 minutes—to do the job. In raising the 6000 pounds 15 feet he did 90,000 foot-pounds of work. Remember—force x distance = work. Since it took him 1800 minutes, he has been working at the rate of 90,000÷1800, or 50 foot-pounds of work per minute.

That's power—the rate of doing work. Thus, power always includes the time element. Doubtless you could do the same amount of work in one 10-hour day, or 600 minutes—which would mean that you would work at the rate of 90,000÷600 = 150 foot-pounds per minute. You then would have a power value three times as great as that of "Lightning."

By formula—

$$\text{Power} = \frac{\text{Work, in ft-lb}}{\text{Time, in minutes}}$$

HORSEPOWER

You measure force in pounds; distance in feet; work in foot-pounds. What is the common unit used for measuring power? The horsepower. If you want to tell someone how powerful an engine is, you could say that it is so many times more powerful than a man, or an ox, or a horse. But what man, and whose ox or horse? James Watt, the fellow who invented the steam engine, compared his early models with the horse. By experiment, he found that an average horse could lift a 330-pound load straight up through a distance of 100 feet in one minute. Figure 8-2 shows you the type of rig he used to find this out. By agreement among scientists, that figure of 33,000 foot-pounds of work done in one minute has been

accepted as the standard unit of power, and it is called a horsepower—hp.

Since there are 60 seconds in a minute, one horsepower is also equal to $\frac{33,000}{60}$ = 550 foot-pounds per second. By formula—

$$\text{Horsepower} = \frac{\text{Power (in ft-lb per min)}}{33,000}$$

CALCULATING POWER

It isn't difficult to figure how much power is needed to do a certain job in a given length of time, nor to predict what size engine or motor is

131.41

Figure 8-1.—Get a horse.

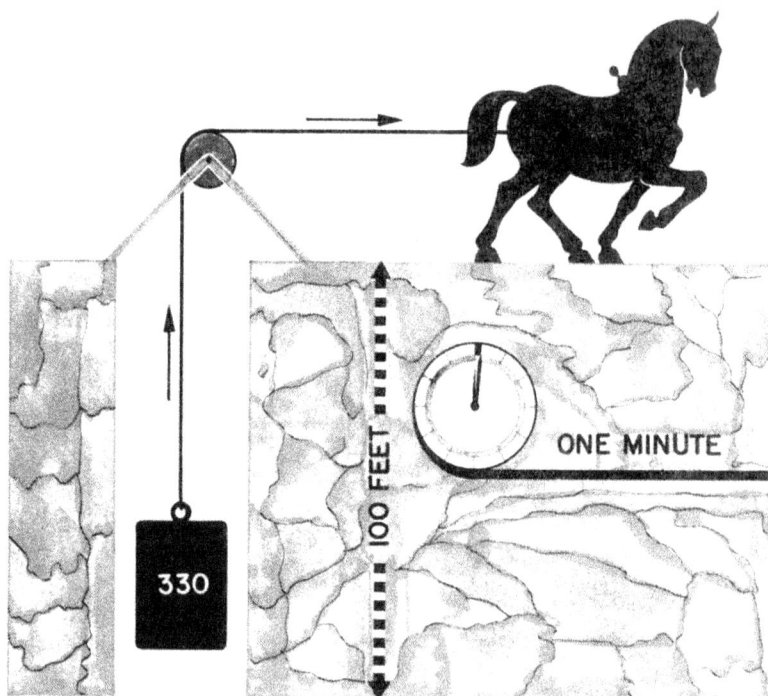

131.42

Figure 8-2.—One horsepower.

needed to do it. Suppose an anchor winch must raise a 6,600-pound anchor through 120 feet in 2 minutes. What must be the theoretical horsepower rating of the motor on the winch?

The first thing to do is to find the rate at which the work must be done. You see the formula—

$$Power \ = \frac{work}{time} \ = \frac{force \ x \ distance}{time}$$

Substitute the known values in the formula, and you get—

$$Power \ = \frac{6,600 \ x \ 120}{2} \ = 396,000 \ ft\text{-}lb/min$$

So far, you know that the winch must work at a rate of 396,000 ft-lb/min. To change this rate to horsepower, you divide by the rate at

which the average horse can work—33,000 ft-lb/min.

$$Horsepower \ = \frac{Power \ (ft\text{-}lb/min)}{33,000} \ =$$

$$= \frac{396,000}{33,000} \ = 12 \ hp.$$

Theoretically, the winch would have to be able to work at a rate of 12 horsepower in order to get the anchor raised in 2 minutes. Of course, you've left out all friction in this problem, so the winch motor would actually have to be larger than 12 hp.

Planes are raised from the hangar deck to the flight deck of a carrier on an elevator. Some place along the line, an engineer had to figure out how powerful the motor had to be in order

47

to raise the elevator. It's not too tough when you know how. Allow a weight of 10 tons for the elevator, and 5 tons for the plane. Suppose that you want to raise the elevator and plane 25 feet in 10 seconds. And that the overall efficiency of the elevator mechanism is 70 percent. With that information you can figure what the delivery horsepower of the motor must be. Set up the formulas—

$$\text{Power} = \frac{\text{force x distance}}{\text{time}}$$

$$\text{hp} = \frac{\text{power}}{33,000}$$

Substitute the known values in their proper places, and you have—

$$\text{power} = \frac{30,000 \text{ x } 25 \text{ ft}}{10/60 \text{ minute}} = 4,500,000 \text{ ft lb/min.}$$

$$\text{hp} = \frac{4,500,000}{33,000} = 136.4 \text{ hp.}$$

So, 136.4 horsepower would be needed if the engine had 100 percent overall efficiency. You want to use 70 percent efficiency, so you use the formula—

$$\text{Efficiency} = \frac{\text{Output}}{\text{Input}}$$

$$\text{Input} = \frac{136.4}{0.70} = 194.8 \text{ hp.}$$

This is the rate at which the engine must be able to work. To be on the safe side, you'd probably select a 200-horsepower auxiliary to do the job.

FIGURING THE HORSEPOWER RATING OF A MOTOR

You have probably seen the horsepower rating plates on electric motors. A number of methods may be used to determine this rating. One way that the rating of a motor or a steam or gas engine can be found is by the use of the prony brake. Figure 8-3 shows you the Prony brake setup. A pulley wheel is fixed to the shaft of the motor, and a leather belt is held firmly against the pulley. Attached to the two ends of the belt are spring scales. When the motor is standing still, each scale reads the same—say 15 pounds.

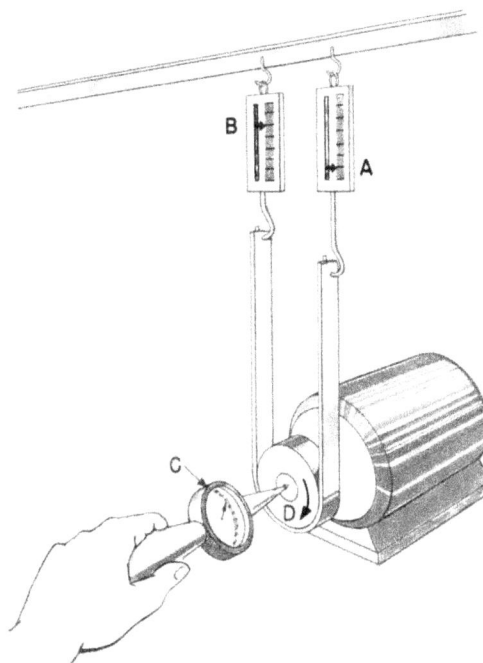

131.43

Figure 8-3.—A prony brake.

When the pulley turns in a clockwise direction, the friction between the belt and the pulley makes the belt try to move with the pulley. Therefore, pull on scale A will be greater, and the pull on scale B will be less than 15 pounds.

Suppose that scale A reads 25 pounds, and scale B reads 5 pounds. That tells you that the drag, or the force against which the motor is working, is 25−5 = 20 pounds. In this case the normal speed of the motor is 1800 rpm (revolutions per minute) and the diameter of the pulley is one foot.

The number of revolutions can be found by holding the revolution counter C against the end of the shaft for one minute. This counter will record the number of turns the shaft makes per minute. The distance D which any point on the pulley travels in one minute is equal to the circumference of the pulley times the number of revolutions—3.14x1x1800 = 5652 ft.

You know that the motor is exerting a force of 20 pounds through that distance. The work done in one minute is equal to the force times the distance, or work = F x D = 20 x 5, 652 = 113,040 ft -lb/min. Change this to horsepower—

$$\frac{113,040}{33,000} = 3.43 \text{ hp.}$$

Here are a few ratings for motors or engines with which you are familar—an electric mixer has a 1/16-hp motor; a washing machine a 1/4-hp motor.

SUMMARY

There are two important points for you to remember about Power—

Power is the rate at which work is done.

The unit in which power is measured is the horsepower, which is equivalent to working at a rate of 33,000 ft-lb per min, or 550 ft-lb per sec.

CHAPTER 9

FORCE AND PRESSURE

By this time you should have a pretty good idea of what a force is. A force is a push or a pull exerted on—or by—an object. You apply a force on a machine, and the machine in turn transmits a force to the load. Men and machines, however, are not the only things that can exert forces. If you've been out in a sailboat you know that the wind can exert a force. Further, you don't have to get knocked on your ear more than a couple of times by the waves to get the idea that water, too, can exert a force. As a matter of fact, from reveille to taps you are almost constantly either exerting forces or resisting them.

MEASURING FORCES

You've had a lot of experience in measuring forces. You can estimate or "guess" the weight of a package you're going to mail by "hefting" it. Or you can put it on a scale to find its weight accurately. Weight is a common term that tells you how much force or pull gravity is exerting on the object.

You can readily measure force with a spring scale. An Englishman named Hooke discovered that if you hang a 1-pound weight on a spring, the spring stretches a certain distance. A 2-pound weight will extend the spring just twice as far, and 3 pounds will lengthen it three times as far as the 1-pound weight did. All you need to do is attach a pointer to the spring, put a face on the scale, and mark on the face the positions of the pointer for various loads in pounds or ounces.

This type of scale can be used to measure the pull of gravity—the weight—of an object, or the force of a pull exerted against friction, as shown in figure 9-1. Unfortunately, springs get tired, just as you do. When they get old, they don't always snap back to the original position. Hence an old spring or an overloaded spring will give inaccurate readings.

HONEST WEIGHT—NO SPRINGS

Because springs do get tired, other types of force-measuring devices are made. You've seen the sign, "Honest Weight—No Springs", on the butchershop scales. Scales of this type are shown in figure 9-2. They are applications of first-class levers. The one shown in figure 9-2A is the simplest type. Since the distance from the fulcrum to the center of each platform is equal, the scale is balanced when equal weights are placed on the platforms. With your knowledge of levers, you will be able to figure out how the steel yard shown in figure 9-2B operates.

PRESSURE

Have you ever tried to walk on crusted snow that would break through when you put your weight on it? But you could walk on the same snow if you put on snowshoes. Further, you know that snowshoes do not reduce your weight—they merely distribute it over a larger area. In doing this, they reduce the pressure per square inch. Figure out how that works. If you weigh 160 pounds, that weight, or force, is more or less evenly distributed by the soles of your shoes. The area of the soles of an average man's shoes is roughly 60 square inches. Each one of those square inches has to carry 160÷60=2.6 pounds of your weight. Since 2.6 pounds per square inch is too much for the snow crust, you break through.

When you put on the snowshoes, you distribute your weight over an area of approximately 900 sq in.—depending, of course, on the size of the snowshoes. Now the force on each one of those square inches is equal to only 160÷900=0.18 pound. The pressure on the snow has been decreased, and the snow can easily support you.

Pressure is force per unit area—and is measured, in pounds per square inch—"psi." With snowshoes on, you exert a pressure of 0.18

131.44

Figure 9-1.—You can measure force with a scale.

psi. To calculate pressure, divide the force by the area over which the force is applied. The formula is—

$$\text{Pressure, in psi} = \frac{\text{Force, in lb}}{\text{Area, in sq in}}$$

Or $$P = \frac{F}{A}$$

To get this idea, follow this problem. A tank for holding fresh water aboard a ship is 10 feet long, 6 feet wide, and 4 feet deep. It holds, therefore, 10x6x4, or 240 cubic feet of water. Each cubic foot of water weighs about 62.5 pounds. The total force tending to push the bottom out of the tank is equal to the weight of the water—240x62.5, or 15,000 lb. What is the pressure on the bottom? Since the weight is evenly distributed on the bottom, you apply the formula $P = \frac{F}{A}$ and substitute the proper values for F and A. In this case, F=15,000 lb, and the area of the bottom in

square inches is 10x6x144, since 144 sq in.=1 sq ft.

$$P = \frac{15,000}{10 \times 6 \times 144} = 1.74 \text{ psi}$$

Now work out the idea in reverse. You live at the bottom of the great sea of air which surrounds the earth. Because the air has weight—gravity pulls on the air, too—the air exerts a force on every object which it surrounds. Near sea level that force on an area of 1 square inch is roughly 15 pounds. Thus, the air-pressure at sea level is about 15 psi. The pressure gets less and less as you go up to higher altitudes.

With your finger, mark out an area of one square foot on your chest. What is the total force which tends to push in your chest? Again use the formula $P = \frac{F}{A}$. Now substitute 15 psi for P, and for A use 144 sq. in. Then F = 144x15, or 2160 lb. The force on your chest is 2160 lb per square foot—more than a ton pushing against an area of

51

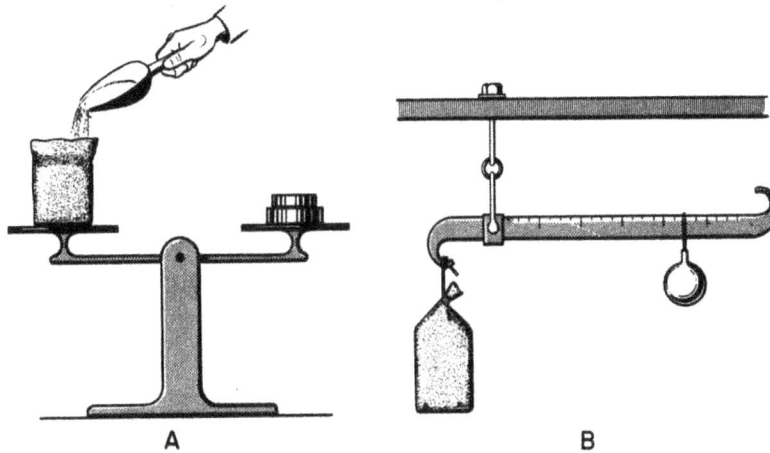

18.30

Figure 9-2.—Balances.

MEASURING PRESSURE

1 sq ft. If there were no air inside your chest to push outward with the same pressure, you'd be squashed flatter than a bride's biscuit.

Fluids—which include both liquids and gases—exert pressure. A fluid at rest exerts equal pressure in all directions. Figure 9-3 shows that. Whether the hole is in the top, the bottom, or in one of the sides of a submarine, the water pushes in through the hole.

In many jobs aboard ship, it is necessary to know the pressure exerted by gas or a liquid. For example, it is important at all times to know

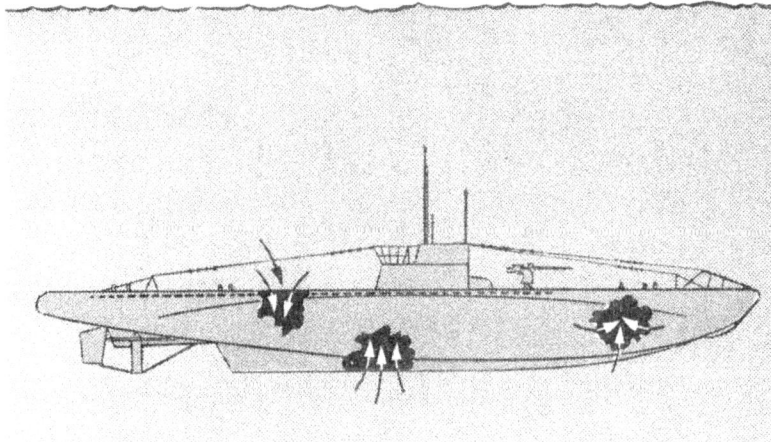

131.45

Figure 9-3.—Fluids exert pressure in all directions.

the steam pressure inside of a boiler. One device to measure pressure is the Bourdon gage, shown in figure 9-4. Its working principle is the same as that of those snakelike paper tubes which you get at a New Year's party. They straighten out when you blow into them.

In the Bourdon gage there is a thin-walled metal tube, somewhat flattened, and bent into the form of a C. Attached to its free end is a lever system which magnifies any motion of the free end of the tube. The fixed end of the gauge ends in a fitting which is threaded into the boiler system so that the pressure in the boiler will be transmitted to the tube. Like the paper "snake," the metal tube tends to straighten out when the pressure inside it is increased. As the tube straightens, the pointer is made to move around the dial. The pressure, in psi, may be read directly on the dial.

Air pressure and pressures of steam and other gases, and fluid pressures in hydraulic systems, are generally measured in pounds per square inch. For convenience, however, the pressure exerted by water is commonly measured in pounds per square foot. You'll find more about this in chapter 10.

The Bourdon gage is a highly accurate but rather delicate instrument, and can be very easily damaged. In addition, it develops trouble where pressure fluctuates rapidly. To overcome this, another type of gauge, the Schrader, was developed. The Schrader gauge (fig. 9-5) is not as accurate as the Bourdon, but is sturdily constructed and quite suitable for ordinary hydraulic pressure measurements. It is especially recommended for fluctuating loads. In the Schrader gauge a piston is directly actuated by the liquid pressure to be measured, and moves up a cylinder against the resistance of a spring, carrying a bar or indicator with it over a calibrated scale. In this manner, all levers, gears, cams, and bearings are eliminated, and a sturdy instrument can be constructed.

Where accurate measurements of comparatively slight pressures are desired, a diaphragm type gauge may be used.

Diaphragm gauges give sensitive and reliable indications of small pressure differences. Diaphragm gauges are often used to measure the air pressure in the space between inner and outer boiler casings. In this type of gauge a diaphragm is connected to a pointer through a metal spring and a simple linkage system (fig. 9-6). One side

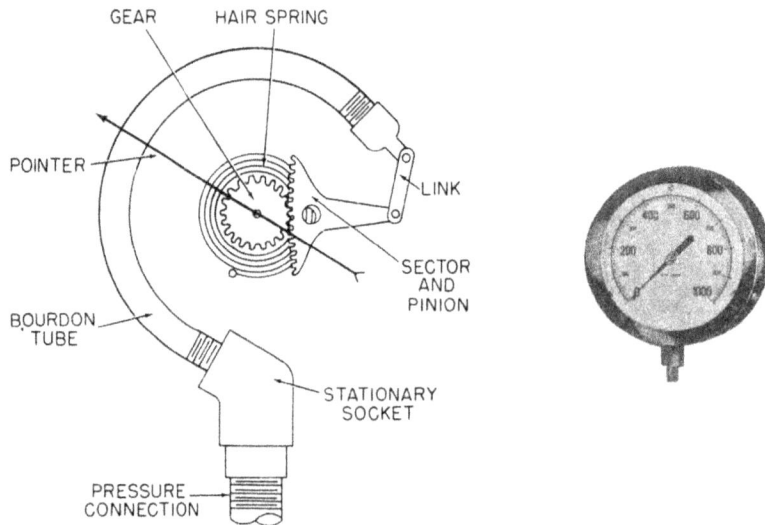

38.211

Figure 9-4.—The Bourdon gauge.

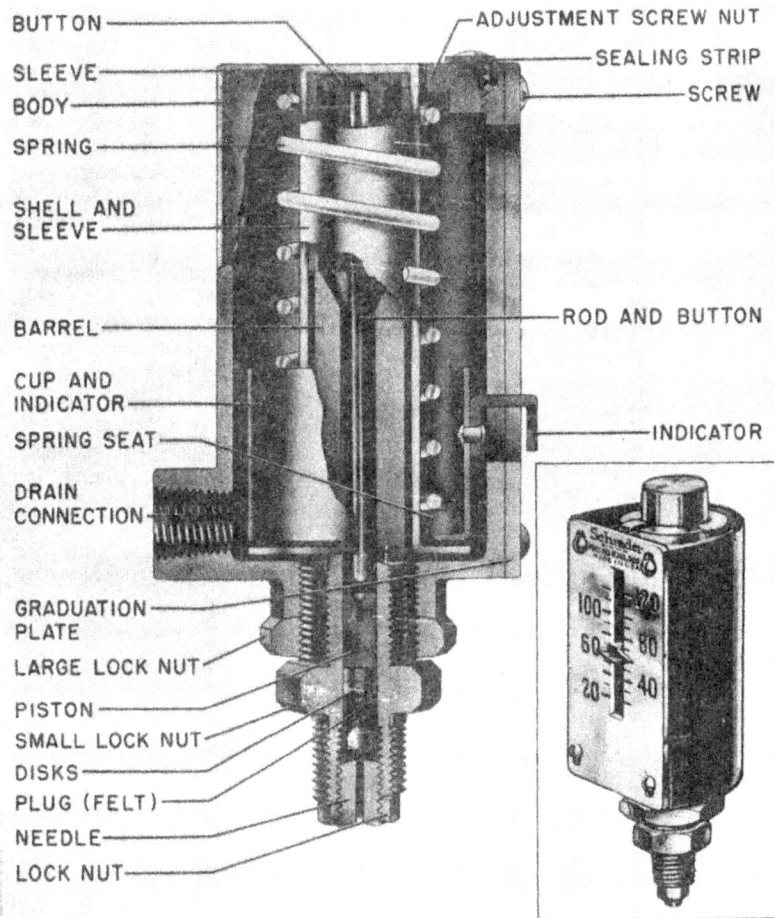

BUTTON — ADJUSTMENT SCREW NUT
SLEEVE — SEALING STRIP
BODY — SCREW
SPRING —
SHELL AND SLEEVE —
BARREL — ROD AND BUTTON
CUP AND INDICATOR —
SPRING SEAT — INDICATOR
DRAIN CONNECTION —
GRADUATION PLATE —
LARGE LOCK NUT —
PISTON —
SMALL LOCK NUT —
DISKS —
PLUG (FELT) —
NEEDLE —
LOCK NUT —

131.46

Figure 9-5.—The Schrader gauge.

of the diaphragm is exposed to the pressure being measured, while the other side is exposed to the pressure of the atmosphere.

Any increase in the pressure line will move the diaphragm upwards against the action of the spring. The linkage system operates and the pointer rotates to a higher reading. When the pressure being measured decreases, the spring moves the diaphragm downward, rotating the pointer to a lower reading. Thus the position of the pointer is a balance between the pressure tending to push the diaphragm upward and the spring action tending to push it down. When the gauge reads "O" the pressure in the line is equal to the outside air pressure.

THE BAROMETER

To the average person the chief importance of weather is as an introduction to general conversation. But at sea and in the air, advance knowledge of what the weather will do is a matter of great concern to all hands. Operations are

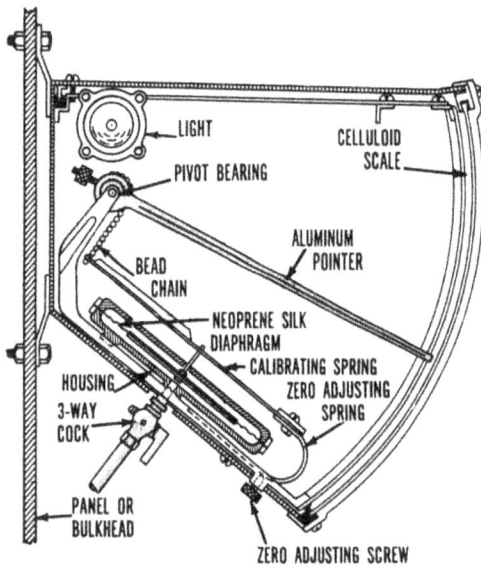

38.212X

Figure 9-6.—Diaphragm pressure gauge.

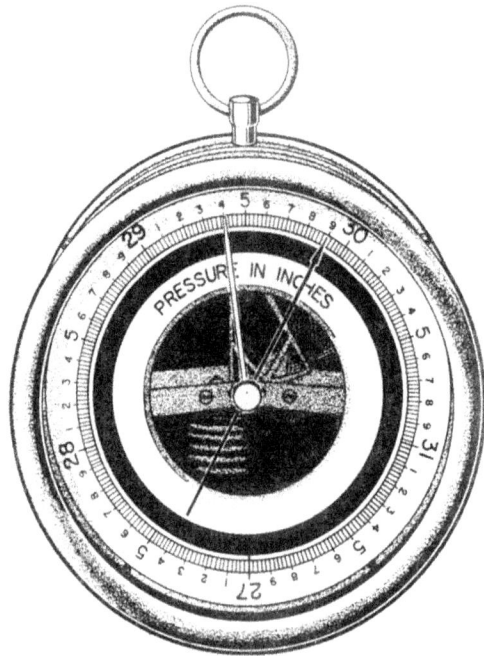

69.87

Figure 9-7.—An aneroid barometer.

planned or cancelled on the basis of weather predictions. Accurate weather forecasts are made only after a great deal of information has been collected by many observers located over a wide area.

One of the instruments used in gathering weather data is the barometer. Remember, the air is pressing on you all the time. So-called normal atmospheric pressure is 14.7 psi. But as the weather changes, the air pressure may be greater or less than normal. If the air pressure is low in the area where you are, you know that air from one or more of the surrounding high-pressure areas is going to move in toward you. Moving air—or wind—is one of the most important factors in weather changes. In general, if you're in a low-pressure area you may expect wind, rain, and storms. A high-pressure area generally enjoys clear weather. The barometer can tell you the air pressure in your locality, and give you a rough idea of what kind of weather may be expected.

The aneroid barometer shown in figure 9-7 is an instrument which measures air pressure. It contains a thin-walled metal box from which

most of the air has been pumped. A pointer is mechanically connected to the box by a lever system. If the pressure of the atmosphere increases, it tends to squeeze in the sides of the box. This squeeze causes the pointer to move towards the high-pressure end of the scale. If the pressure decreases, the sides of the box expand outward. This causes the pointer to move toward the low-pressure end of the dial.

Notice that the numbers on the dial run from 27 to 31. To understand why these particular numbers are used, you have to understand the operation of the mercurial barometer. You see one of these in figure 9-8. It consists of a glass tube partly filled with mercury. The upper end is closed. There is a vacuum above the mercury in the tube, and the lower end of the tube is submerged in a pool of mercury in an open cup. The atmosphere presses down on the mercury in the cup, and tends to push the mercury up in the tube. The greater the air pressure, the higher the column of mercury rises. At sea level, the

55

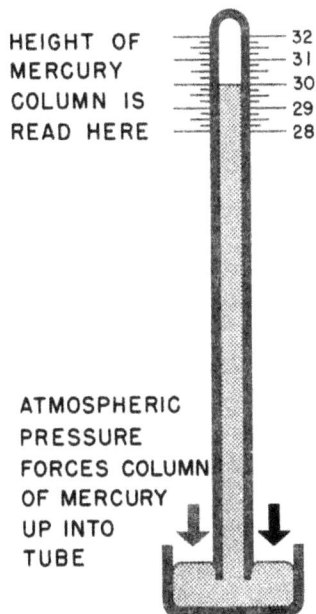

HEIGHT OF
MERCURY
COLUMN IS
READ HERE

— 32
— 31
— 30
— 29
— 28

ATMOSPHERIC
PRESSURE
FORCES COLUMN
OF MERCURY
UP INTO
TUBE

69.86

Figure 9-8.—A mercurial barometer.

normal pressure is 14.7 psi, and the height of the mercury in the tube is 30 inches. As the air pressure increases or decreases from day to day, the height of the mercury rises or falls. A mercury barometer aboard ship is usually mounted in gimbals to keep it in a vertical position despite the rolling and pitching of the ship.

Pressures indicated on dials of most gauges are relative. That is, they are either greater or less than normal. But remember—the dial of an aneroid barometer always indicates absolute pressures, not relative. When the pressure exerted by any gas is less than 14.7 psi, you have what's called a partial vacuum. The condensers on steam turbines, for instance, are operated at pressure well below 14.7 psi. Steam under very high pressure is run into the turbine and causes the rotor to turn. After it has passed through the turbine it still exerts a back pressure against the blades. You can see that this is bad. Soon the back pressure would be nearly as

large as that of the incoming steam, and the turbine would not turn at all. To reduce the back pressure as much as possible, the exhaust steam is run through pipes which are surrounded by cold sea water. This causes the steam in the pipes to condense into water, and the pressure drops well below atmospheric pressure.

It is important for the engineer to know the pressure in the condensers at all times. To measure this reduced pressure, or partial vacuum, he uses a gauge called a manometer. Figure 9-9 shows you how this simple device is made. A U-shaped tube has one end connected to the low-pressure condenser and the other end is open to the air. The tube is partly filled with colored water. The normal air pressure on the open end exerts a bigger push on the colored water than the push of the low-pressure steam, and the colored water is forced part way up into the left arm of the tube. From the scale between the two arms of the U, the difference in the height of the two columns of water can be read. This tells the engineer the degree of vacuum—or how much below atmospheric pressure the pressure is in the condenser.

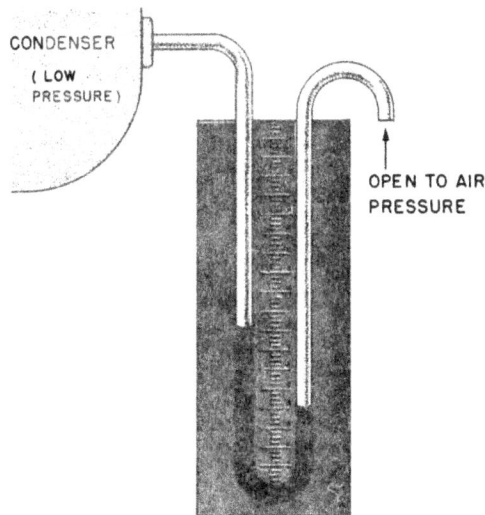

CONDENSER
(LOW
PRESSURE)

OPEN TO AIR
PRESSURE

Figure 9-9.—A manometer.

SUMMARY

Here are seven points that you should remember—
- A force is a push or a pull exerted on—or by—an object.
- Force is generally measured in pounds.
- Pressure is the force per unit area which is exerted on, or by, an object. It is commonly measured in pounds per square inch—psi.
- Pressure is calculated by the formula $P = \frac{F}{A}$.
- Spring scales and lever balances are familiar instruments you use for measuring forces. Bourdon gauges, barometers, and manometers are instruments for the measurement of pressure.
- The normal pressure of the air is 14.7 psi at sea level.
- Pressure is generally relative. It is sometimes greater—sometimes less—than normal air pressure. When pressure is less than the normal air pressure, you call it vacuum.

CHAPTER 10

HYDROSTATIC AND HYDRAULIC MACHINES

HYDROSTATIC MACHINES

LIQUIDS AT REST

You know that liquids exert pressure. In order that your ship may remain afloat, the water must push upward on the hull. But the water is also exerting pressure on the sides. If you are billeted on a submarine, you are more conscious of water pressure—when you're submerged the sub is being squeezed from all sides. If your duties include deep-sea diving, you'll go over the side pumped up like a tire so that you can withstand the terrific force of the water below. The pressure exerted by the sea water, or by any liquid at rest, is called hydrostatic pressure. In handling torpedoes, mines, depth charges, and some types of aerial bombs, you'll be dealing with devices which are operated by hydrostatic pressure.

In chapter 9, you found out that all fluids exert pressure in all directions. That's simple enough. But how great is the pressure? Try a little experiment. Place a pile of blocks in front of you on the table. Stick the tip of your finger under the first block from the top. Not much pressure on your finger, is there? Stick it in between the third and fourth blocks. The pressure on your finger has increased. Now slide your finger under the bottom block in the pile. There you find the pressure is greatest. The pressure increases as you go lower in the pile. You might say that pressure increases with depth. The same is true in liquids. The deeper you go, the greater the pressure becomes. But, depth isn't the whole story.

Suppose the blocks in the preceding paragraph were made of lead. The pressure at any level in the pile would be considerably greater. Or, suppose they were blocks of balsa wood—the

pressure at each level wouldn't be so great. Pressure, then, depends not only on the depth, but also on the weight of the material. Since you are dealing with pressure—force per unit of area—you will also be dealing with weight per unit of volume—or density.

When you talk about the density of a substance you are talking about its weight per cubic foot—or per cubic inch. For example, the density of water is 62.5 lb. per cu. ft. This gives you a more exact way of comparing the weights of two materials. To say that lead is heavier than water isn't a complete statement. A 22-caliber bullet doesn't weigh as much as a pail of water. It is true, however, that a cubic foot of lead is lots heavier than a cubic foot of water. Lead has a greater density than water. The density of lead is 710 lb. per cu. ft., as compared with 62.5 lb. per cu. ft. for water.

Pressure depends on two factors—depth and density—so it is easy to write a formula that will help you find the pressure at any depth in any liquid. Here it is—

$$P = H \times D$$

in which

P = pressure, in lb. per sq. in., or lb. per sq. ft

H = depth of the point, measured in feet or inches.

and

D = density in lb. per cu. in. or in lb. per cu. ft.

Note: If inches are used, they must be used throughout; if feet are used, they must be used throughout.

What is the pressure on one square foot of the surface of a submarine if the submarine is 200 feet below the surface? Use the formula—

$$P = H \times D$$

and

$$P = 200 \times 62.5 = 12,500 \text{ lb. per sq. ft.}$$

Every square foot of the sub's surface which is at that depth has a force of over 6 tons pushing in on it. If the height of the hull is 20 feet, and the area in question is midway between the sub's top and bottom, you can see that the pressure on the hull will be at least (200—10) x 62.5 = 11,875 lb. per sq. ft., and the greatest pressure will be (200 + 10) x 62.5 = 13,125 lb. per sq. ft. Obviously, the hull has to be made very strong to withstand such pressures.

Using Pressure to Fire
the Depth Charge

Although hiding below the surface exposes the sub to great fluid pressure, it also provides the sub with a great advantage. A submarine is hard to kill because it is hard to hit. A depth charge must explode within 30 to 50 feet of a submarine to really score. And that means the depth charge must not go off until it has had time to sink to approximately the same level as the sub. You use a firing mechanism which is set off by the pressure at the estimated depth of the submarine.

Figure 10-1 shows a depth charge and its interior components. A depth charge is a sheet-metal container filled with a high explosive and a firing device. A tube passes through its center from end to end. Fitted in one end of this tube is the booster, which is a load of granular TNT to set off the main charge. The safety fork is knocked off on launching, and the inlet valve cover is removed from an inlet through which the water enters.

When the depth charge gets about 12 to 15 feet below the surface, the water pressure is sufficient to extend a bellows in the booster extender. The bellows trips a release mechanism, and a spring pushes the booster up against the centering flange. Notice that the detonator fits into a pocket in the booster. Unless the detonator is in this pocket, it cannot set off the booster charge.

Nothing further happens until the detonator is fired. As you can see, the detonator is held in the end of the pistol, with the firing pin aimed

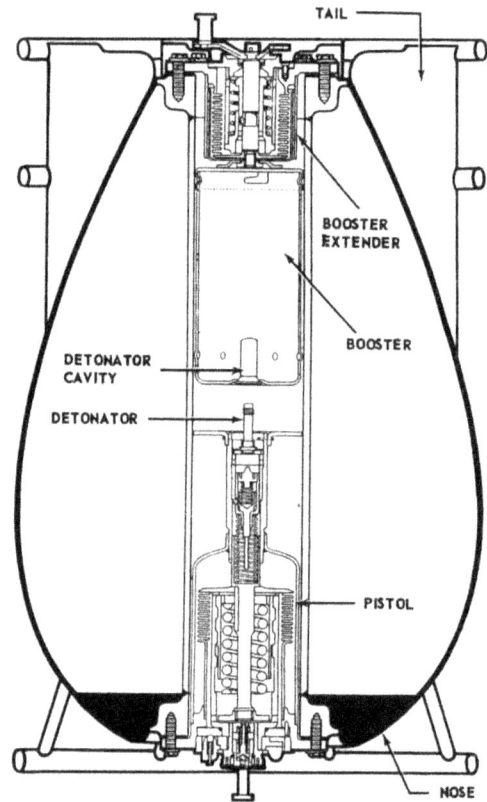

4.198

Figure 10-1.—A depth charge.

at the detonator base. The pistol also contains a bellows into which the water rushes as the charge goes down. As the pressure increases, the bellows begins to expand against the depth spring. You can adjust this spring so that the bellows will have to exert a predetermined force in order to compress it. Figure 10-2 shows you the depth-setting dials of one type of depth charge. Since the pressure on the bellows depends directly on the depth, you can arrange to have the charge go off at any depth you select on the dial. When the pressure in the bellows becomes sufficiently great it releases the firing spring, which drives the firing pin into the detonator. The booster, already moved into position, is fired, and this in turn sets off the entire load of TNT.

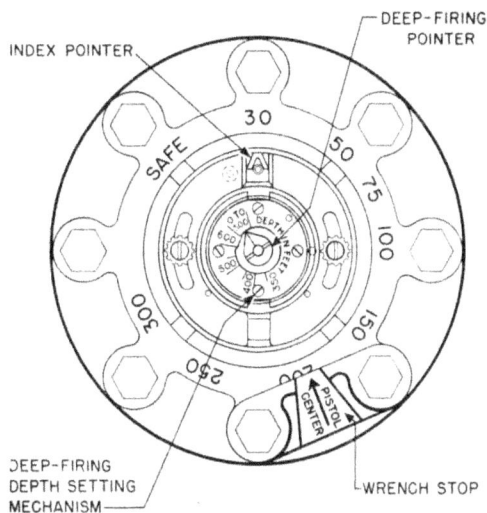

4.204

Figure 10-2.—You select the depth on these dials.

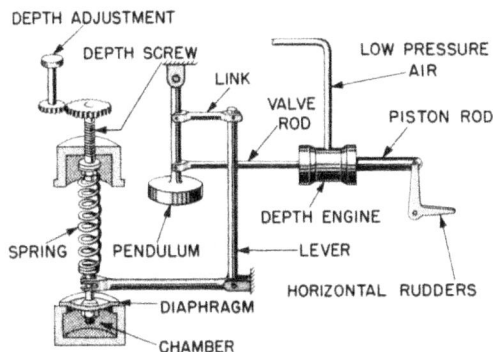

4.122

Figure 10-3.—Inside a torpedo.

These two bellows—operated by hydrostatic pressure—serve two purposes. First they permit the depth charge to be fired at the proper depth; second, they make the charge safe to handle and carry. If the safety fork and the valve inlet cover should accidentally be knocked off on deck, nothing would happen. Even if the detonator went off while the charge was being handled, the main charge would not let go unless the booster were in the extended position.

To keep a torpedo on course toward its target is quite a job. Maintaining the proper compass course by the use of a gyroscope is only part of the problem. The torpedo must travel at the proper depth so that it will neither pass under the target ship nor hop out of the water on the way. Here again hydrostatic pressure is used to advantage.

As figure 10-3 indicates, the tin fish contains an airfilled chamber which is sealed with a thin, flexible metal plate, or diaphragm. This diaphragm can bend upward or downward against the spring. The tension on this spring is determined by setting the depth-adjusting knob.

Suppose the torpedo starts to dive below the selected depth. The water, which enters the torpedo and surrounds the chamber, exerts an increased pressure on the diaphragm and causes it

to bend down. If you follow the lever system, you can see that the pendulum will be pushed forward. Notice that a valve rod connects the pendulum to the piston of the depth engine. As the piston moves to the left, low-pressure air from the torpedo's air supply enters the depth engine to the right of the piston and pushes it to the left. A depth engine must be used because the diaphragm is not strong enough to move the rudders.

The depth-engine's piston is connected to the horizontal rudders as shown. When the piston moves to the left, the rudder is turned upward, and the torpedo begins to rise to the proper depth. If the nose goes up, the pendulum tends to swing backward and keep the rudder from elevating the torpedo too rapidly. As long as the torpedo runs at the selected depth, the pressure on the chamber remains constant, and the rudders do not change from their horizontal position.

Pressure and the
Deep-Sea Diver

Navy divers have a practical, first-hand knowledge of hydrostatic pressure. Think what happens to a diver who goes down 100 feet to work on a salvage job. The pressure on him at that depth is 6,250 lbs. per sq. ft.! Something must be done about that, or he'd be squashed flatter than a pancake.

To counterbalance this external pressure, the diver is enclosed in a rubber suit into which air under pressure is pumped by a shipboard compressor. Fortunately, the air not only inflates the suit, but gets inside of the diver's body as well. It enters his lungs, and even gets into his blood stream which carries it to every part of his body. In that way his internal pressure can be kept just equal to the hydrostatic pressure.

As he goes deeper, the air pressure is increased to meet that of the water. In coming up, the pressure on the air is gradually reduced. If he is brought up too rapidly, he gets the "bends." The air which was dissolved in his blood begins to come out of solution, and form as bubbles in his veins. Any sudden release in the pressure on a fluid results in freeing some of the gases which are dissolved in the fluid.

You have seen this happen when you suddenly relieve the pressure on a bottle of pop by removing the cap. The careful matching of hydrostatic pressure on the diver by means of air pressure in his suit is essential if diving is to be done at all.

A Sea-Going Speedometer

Here's another device that shows you how your Navy applies its knowledge of hydrostatic pressure. Did you ever wonder how the skipper knows the speed the ship is making through the water? There are several instruments used to give this information—the patent log, the engine revolution counter, and the pitometer log. The "PIT. LOG" is operated, in part, by hydrostatic pressure. It really indicates the difference between hydrostatic pressure and the pressure of the water flowing past the ship—but you can use this difference to indicate ship's speed.

Figure 10-4 shows you a schematic drawing of a pitometer log. A double-wall tube sticks out forward of the ship's hull into water which is not disturbed by the ship's motion. In the tip of the tube is an opening A. When the ship is moving there are two forces or pressures acting on this opening—the hydrostatic pressure due to the depth of water above the opening, and a pressure caused by the push of the ship through the water. The total pressure from these two forces is transmitted through the central or white tube to the left-hand arm of a manometer.

In the side of the tube is a second opening B which does not face in the direction the ship is moving. Opening B passed through the outer wall of the double-wall tube, but not through the inner wall. The only pressure affecting this opening

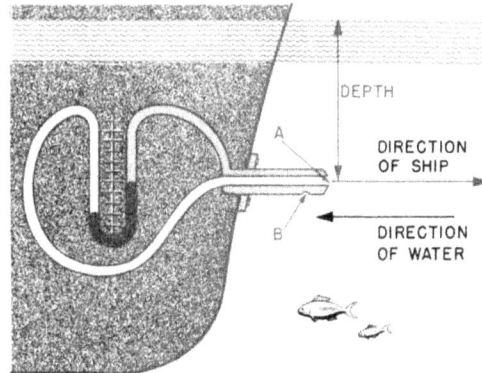

131.47

Figure 10-4.—A Pitometer log.

B is the hydrostatic pressure. This pressure is transmitted through the outer tube (shaded in the drawing) to the right-hand arm of the manometer.

When the ship is dead in the water, the pressure through both openings A and B is the same, and the mercury in each arm of the manometer stands at the same level. However, as soon as the ship begins to move, additional pressure is developed at opening A, and the mercury is pushed down in the left-hand arm and up into the right-hand arm of the tube. The faster the ship goes, the greater this additional pressure becomes, and the greater the difference will be between the levels of the mercury in the two arms of the manometer. The speed of the ship can be read directly from the calibrated scale on the manometer.

Incidentally—since air is also a fluid—the airspeed of an aircraft can be found by a similar device. You have probably seen the thin tube sticking out from the leading edge of a wing, or from the nose of the plane. Flyers call this tube a pitot tube. Its fundamental principle is the same as that of the pitometer log.

SUMMARY

The Navy uses many devices whose operation is dependent on the hydrostatic principle. Here are three points to remember about the operation of these devices.

Pressure in a liquid is exerted equally in all directions.

You use the term hydrostatic pressure when you are talking about the pressure at any depth in a liquid that is not flowing. Pressure depends upon both depth and density.

The formula for finding pressure is—

$$P = H \times D$$

HYDRAULIC MACHINES

LIQUIDS IN MOTION

Perhaps your earliest contact with a hydraulic machine was when you got your first haircut. Tony put a board across the arms of the chair, sat you on it, and began to pump the chair up to a convenient level. As you grew older, you probably discovered that the filling station attendant could put a car on the greasing rack, and—by some mysterious arrangement— jack it head-high. No doubt the attendant told you that oil under pressure below the piston was doing the job.

Come to think about it, you've probably known something about hydraulics for a long time. Automobiles and airplanes use hydraulic brakes. As a bluejacket, you'll have to operate many hydraulic machines, so you'll want to understand the basic principles on which they work.

Simple machines such as the lever, the inclined plane, the pulley, the wedge, and the wheel and axle, were used by primitive man. But it was considerably later before someone discovered that liquids and gases could be used to exert forces at a distance. Then, a vast number of new machines appeared. A machine which transmits forces by means of a liquid is a hydraulic machine. A variation of the hydraulic machine is the type that operates by the use of a compressed gas. This type is called the pneumatic machine. This chapter deals only with basic hydraulic machines.

Pascal's Law

A Frenchman named Pascal discovered that a pressure applied to any part of a confined fluid is transmitted to every other part with no loss. The pressure acts with equal force on all equal areas of the confining walls, and perpendicular to the walls.

But remember this—when you are talking about the hydraulic principle as applied to a hydraulic machine, you are talking about the way a liquid acts in a closed system of pipes and cylinders. The action of a liquid under such conditions is somewhat different from its behavior in open containers, or in lakes, rivers, or oceans. You should also keep in mind that most liquids cannot be compressed—squeezed into a smaller space. Liquids don't "give" the way air does when pressure is applied, nor do liquids expand when pressure is removed.

Punch a hole in a tube of shaving cream. If you push down at any point on the tube the cream comes out of the hole. Your force has been transmitted from one place to another by the shaving cream—which is fluid—a thick liquid. Figure 10-5 shows what would happen if you punched four holes in the tube. If you press on the tube at one point, the cream comes out of all four holes. This tells you that a force applied on a liquid is transmitted equally in every direction to all parts of the container. Right there you have illustrated a basic principle of hydraulic machines.

This principle is used in the operation of four-wheel hydraulic automobile brakes. Figure 10-6 is a simplified drawing of this brake system. You push down on the brake pedal and force the piston in the master cylinder against the fluid in that cylinder. This push sets up a pressure on the fluid just as your finger did on the shaving cream in the tube. The pressure on the fluid in the master cylinder is transmitted through the lines to the brake cylinders in each wheel. This fluid under pressure pushes against the pistons in each of the brake cylinders and forces the brake shoes out against the drums.

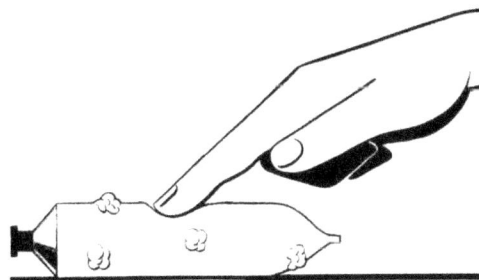

131.48

Figure 10-5.—Pressure is transmitted in all directions.

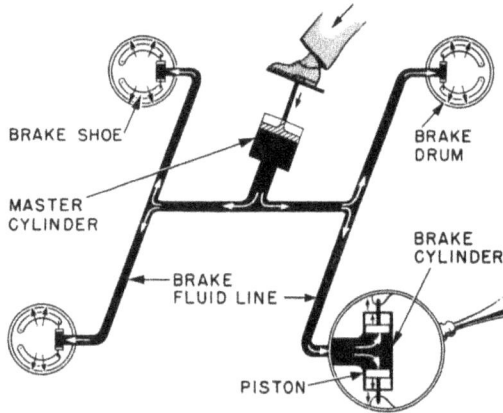

81.280

Figure 10-6.—Hydraulic brakes.

Mechanical Advantage
Through Hydraulics

The next thing to understand about hydraulic machines is the relationship between the force you apply and the result you get. Figure 10-7 will help you on this. The U-shaped tube has a cross-sectional area of one sq. inch. In each arm there's a piston which fits snugly, but which can move up and down. If you place a one-pound weight on one piston, the other will be pushed out the top of its arm immediately. Place a one-pound weight on each piston, however, and they remain in their original positions, as shown in figure 10-8.

Thus you see that a pressure of one pound per sq. in. applied downward on the right-hand piston exerts a pressure of one pound per sq. in. upward against the left-hand one. In other words, not only is the force transmitted by the liquid around the curve, but the force is the same on each unit area of the container. It makes no difference how long the connecting tube is, or how many turns it makes. It is important, however, that the entire system be full of liquid. Hydraulic systems will fail to operate properly if air is present in the lines or cylinders.

Now look at figure 10-9. The piston on the right has an area of one sq. in., but the piston on the left has an area of 10 sq. in. If you push down on the smaller piston with a force of one pound, the liquid will transmit this pressure to every square inch of surface in the system. Since the left-hand piston has an area of 10 sq. in., and each square inch has a force of one pound transmitted to it, the total effect is to push on the larger piston with a total force of 10 pounds. Set a 10-pound weight on the larger piston and it will be supported by the one-pound force of the smaller piston.

There you have a one-pound push resulting in a 10-pound force. That's a mechanical advantage of ten. This is why hydraulic machines

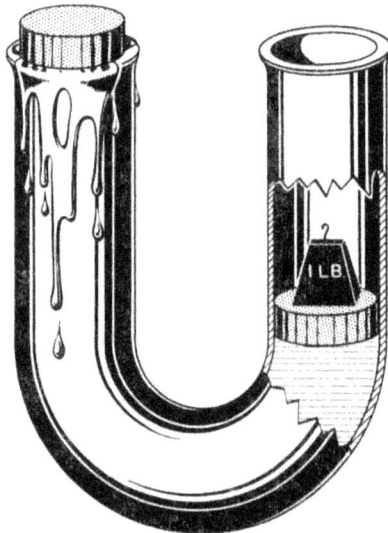

131.49

Figure 10-7.—The liquid transmits the force.

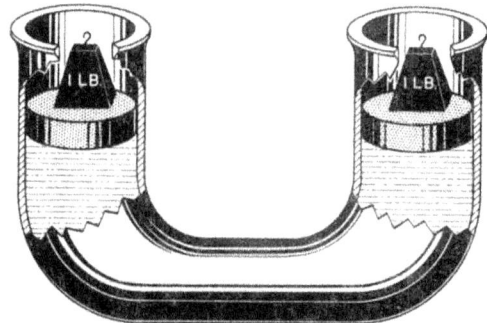

5.181

Figure 10-8.—Pressure is the same on all parts of an enclosed liquid.

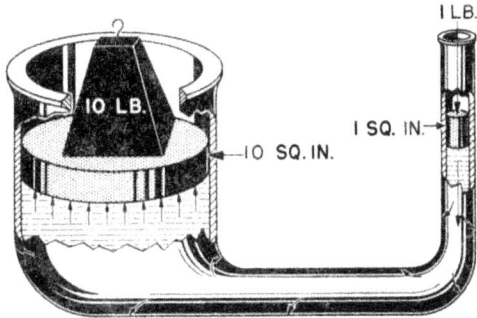

4.7
Figure 10-9.—A mechanical advantage of 10.

are important. Here's a formula which will help you to figure the forces that act in a hydraulic machine—

$$\frac{F_1}{F_2} = \frac{A_1}{A_2}$$

In which F_1 = force, in pounds, applied to the small piston,
F_2 = force, in pounds, applied to the large piston,
A_1 = area of small piston, in square inches,
A_2 = area of large piston, in square inches.

Try out the formula on the hydraulic press in figure 10-10. The large piston has an area of 90 sq. in. and the smaller one an area of two sq. in. The handle exerts a total force of 15 pounds on the small piston. With what total force will the large piston be raised?

Write down the formula—

$$\frac{F_1}{F_2} = \frac{A_1}{A_2}$$

Substitute the known values—

$$\frac{15}{F_2} = \frac{2}{90}$$

and—

$$F_2 = \frac{90 \times 15}{2} = 675 \text{ pounds}.$$

Where's The Catch?

You know from your experience with levers that you can't get something for nothing. Applying this knowledge to the simple system in figure 10-9, you know that you can't get a 10-pound force from a one-pound effort without sacrificing distance. The one-pound effort will have to be applied through a much greater distance than the 10-pound force will move. If you raise the 10-pound weight through a distance of one foot, through what distance will the one-pound effort have to be applied? Remember—if you neglect friction, the work done on any machine equals the work done by that machine. Use the work formula, and you can find how far the smaller piston will have to move.

Work input = Work output

$$F_1 \times D_1 = F_2 \times D_2$$

By substituting— $1 \times D_1 = 10 \times 1$

and— $D_1 = 10 \text{ feet}$

There's the catch. The smaller piston will have to move through a distance of 10 feet in order to raise the 10-pound load one foot. It looks then as though the smaller cylinder would have to be at least 10 feet long—and that wouldn't be practical. Actually, it isn't necessary—if you put a valve in the system.

The hydraulic press in figure 10-10 contains a valve for just this purpose. As the small piston moves down, it forces the fluid past the check valve A into the large cylinder. As soon as you start to move the small piston upward, the pressure to the right of the check valve A is removed, and the pressure of the fluid below the large piston helps the check valve spring force that valve shut. The liquid which has passed through the valve opening on the down stroke of the small piston is trapped in the large cylinder.

The small piston rises on the up-stroke until its bottom passes the opening to the fluid reservoir. More fluid is sucked past a check valve B and into the small cylinder. The next down-stroke forces this new charge of fluid out of the small cylinder past the check valve into the large cylinder. This process is repeated stroke by stroke until enough fluid has been forced into the large cylinder to raise the large piston the required distance of one foot. The force has been applied through a distance of 10 feet on the pump handle,

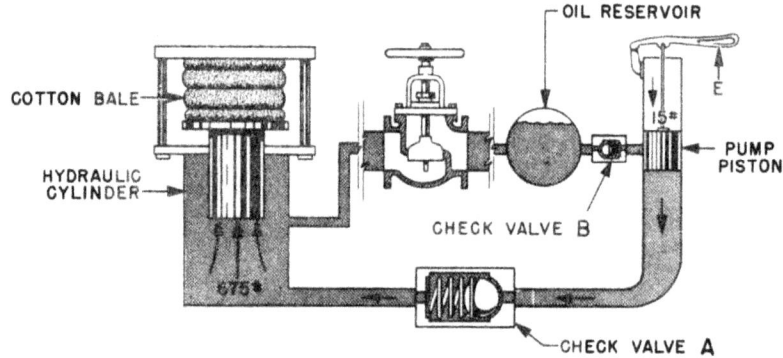

Figure 10-10.—Hydraulic press.

131.50

but it was done by making a series of relatively short strokes—the sum of all the strokes being equal to 10 feet.

Maybe you're beginning to wonder how the large piston gets back down after you've baled the cotton. The fluid can't run back past the check valve B—that's obvious. You lower the piston by letting the oil flow back to the reservoir through a return line. Notice that a simple gate valve is inserted in this line. When the gate valve is opened, the fluid flows back into the reservoir. Of course, this valve is kept shut while the pump is in operation.

Hydraulics Aid the Helmsman

You've probably seen the helmsman swing a ship weighing thousands of tons about as easily as you turn your car. No, he's not a superman. He does it with machines.

Many of these machines are hydraulic. There are several types of hydraulic and electro-hydraulic steering mechanisms, but the simplified diagram in figure 10-11 will help you to understand the general principles of their operation. As the hand steering wheel is turned in a counterclockwise direction, its motion turns the pinion gear g. This causes the left-hand rack r_1 to move downward, and the right-hand rack r_2 to move upward. Notice that each rack is attached to a piston P_1 or P_2. The downward motion of rack r_1 moves piston p_1 downward in its cylinder and pushes the oil out of that cylinder through the line. At the same time, piston p_2 moves upward and pulls oil from the right-hand line into the right-hand cylinder.

If you follow these two lines, you see that they enter a hydraulic cylinder S—one line entering above and one below the single piston in that cylinder. In the direction of the oil flow in the diagram, this piston and the attached plunger are pushed down toward the hydraulic pump h. So far, in this operation, you have used hand power to develop enough oil pressure to move the control plunger attached to the hydraulic pump. At this point an electric motor takes over and drives the pump h.

Oil is pumped under pressure to the two big steering rams R_1 and R_2. You can see that the pistons in these rams are connected directly to the rudder crosshead which controls the position of the rudder. With the pump operating in the direction shown, the ship's rudder is thrown to the left, and the bow will swing to port. This operation demonstrates how a small force applied on the steering wheel sets in motion a series of operations which result in a force of thousands of pounds.

Getting Planes on Deck

The swift, smooth power required to get airplanes from the hanger deck to the flight deck of a carrier is supplied by a hydraulic lift. Figure 10-12 explains how this lifting is done. A variable-speed gear pump is driven by an electric motor. Oil enters the pump from the reservoir and is forced through the lines to four hydraulic rams, the pistons of which raise the elevator platform. The oil under pressure exerts its force on each square inch of surface area of the four pistons. Since the pistons are large, a

65

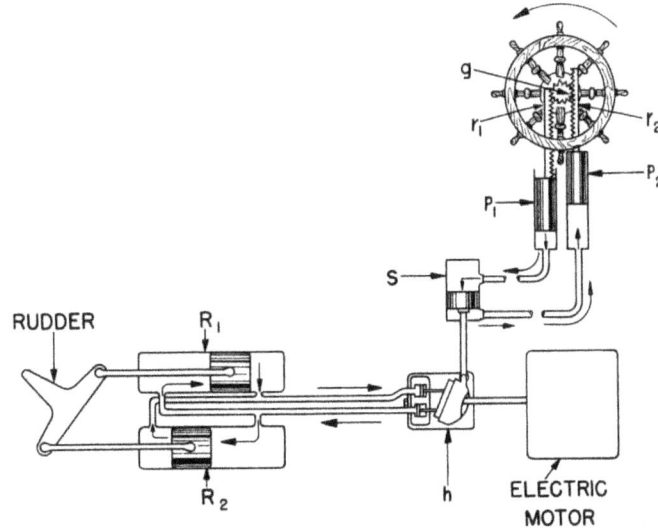

131.51

Figure 10-11.—Steering is easy with this machine.

large total lifting force results. The elevator can be lowered by reversing the pump, or by opening valve 1 and closing valve 2. The weight of the elevator will then force the oil out of the cylinders and back into the reservoir.

Submarines Use Hydraulics

Here's another application of hydraulics which you will find interesting. Inside a submarine, between the outer skin and the pressure hull, are several tanks of various design and purpose. These are used to control the total weight of the ship, allowing it to submerge or surface, and to control the trim, or balance fore and aft, of the submarine. The main ballast tanks have the primary function of either destroying or restoring positive buoyancy in the submarine. By allowing air to escape through hydraulically operated vents at the top of the tanks, sea water is able to enter through the flood ports at the bottom—replacing the air that had been holding it out. To regain positive buoyancy, the tanks are "blown" free of sea water with compressed air. Sufficient air is then left trapped in the tanks to prevent the sea water from reentering.

Other tanks, such as the variable ballast tanks and special ballast tanks like the negative tank, safety tank, and bow buoyancy tank, are used either to control trim, or stability, or for emergency weight compensating purposes. The variable ballast tanks have no direct connection to the sea. Therefore, water must be pumped into or out of them. The negative tank and the safety tank, however, can be opened to the sea through large flood valves. These valves, as well as the vent valves for the main ballast tanks and those for the safety and negative tanks, are all hydraulically operated. The vents and flood valves are outside of the pressure hull, so some means of remote control is necessary if they are to be opened and closed from within the submarine. For this purpose, hydraulic pumps, lines, and rams are used. Oil pumped through tubing running through the pressure hull actuates the valve's operating mechanisms by exerting pressure on and moving a piston in a hydraulic cylinder. It is easier and simpler to operate the valves by a hydraulic system from a control room than it would be to do so by a mechanical system of gears, shafts, and levers. The hydraulic lines can be readily led around corners and obstructions, and a minimum of moving parts is required.

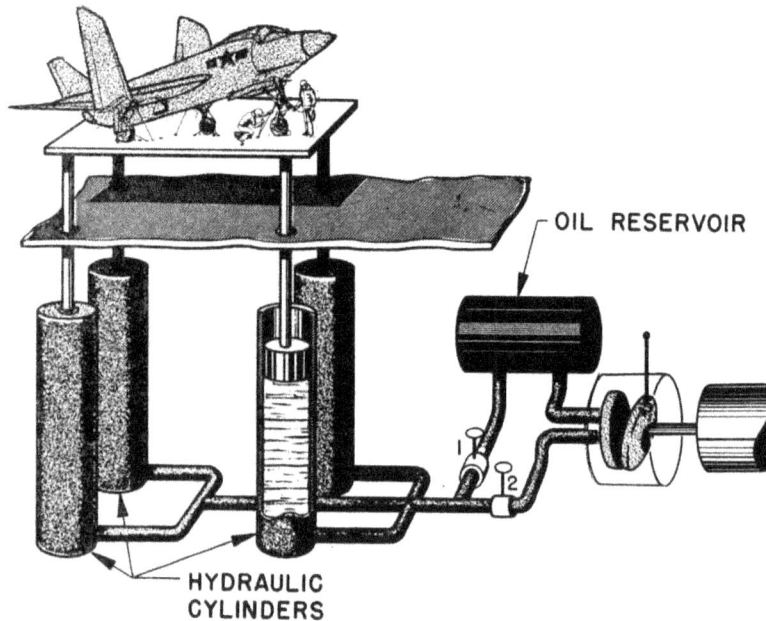

131.52

Figure 10-12.—This gets them there in a hurry.

Figure 10-13 is a schematic sketch of the safety tank—one of the special ballast tanks in a submarine. The main vent and the flood valves of this tank are operated hydraulically from remote control; although, in an emergency, they may be operated manually.

Hydraulics are also used in many other ways aboard the submarine. The periscope is raised and lowered, the submarine is steered, and the bow and stern planes are controlled by means of hydraulic systems. The windlass and capstan system, used in mooring the submarine, is hydraulically operated, and many more applications of hydraulics can be found aboard the submarine.

The Accumulator

In some hydraulic systems, oil is kept under pressure in a container called an accumulator. Figure 11-14 shows you this large cylinder, into the top of which oil is pumped. A free piston

divides the cylinder into two parts. Compressed air is forced in below the piston at a pressure of, say, 600 psi. Oil is then forced in on top of the piston. As the pressure above it increases, the piston is forced down, and squeezes the air into a smaller space. Air is elastic—it can be compressed under pressure—but it will expand as soon as the pressure is reduced. When oil pressure is reduced, relatively large quantities of oil under working pressure are instantly available to operate hydraulic rams or motors any place on the sub.

SUMMARY

The working principle of all hydraulic mechanisms is simple enough. Whenever you find an application that seems a bit hard to understand, keep these points in mind—

Hydraulics is the term applied to the behavior of enclosed liquids. Machines which are operated by liquids under pressure are called hydraulic machines.

67

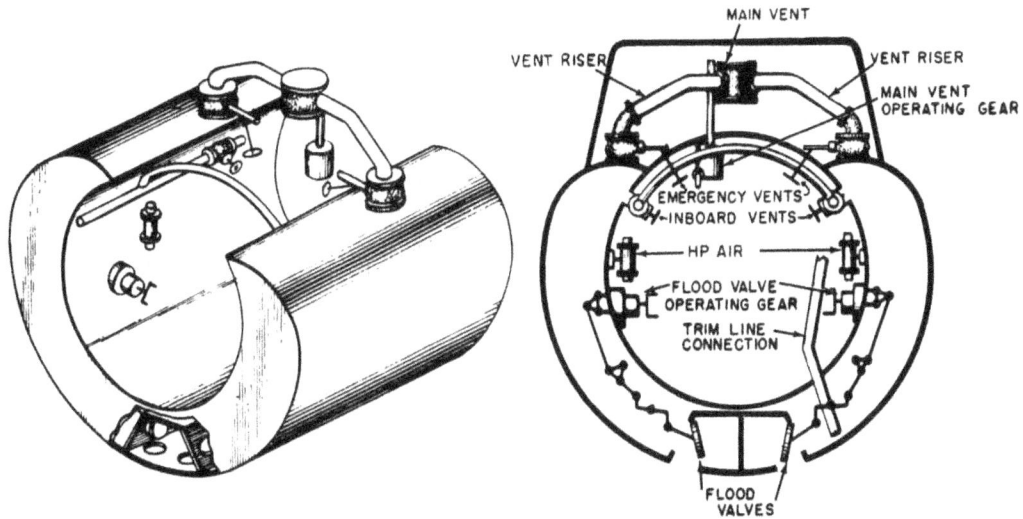

131.53

Figure 10-13.—Submarine special ballast tank (safety tank).

Liquids are incompressible. They cannot be squeezed into spaces smaller than they originally occupied.

A force applied on any area of a confined liquid is transmitted equally to every part of the liquid.

In hydraulic cylinders, the relation between the force exerted by the larger piston to the force applied on the smaller piston is the same as the relation between the area of the larger piston and the area of the smaller piston.

Some of the advantages of hydraulic machines are—

Tubing is used to transmit forces, and tubing can readily transmit forces around corners.

Little space is required for tubing.

Few moving parts are required.

Efficiency is high, generally 80-95%.

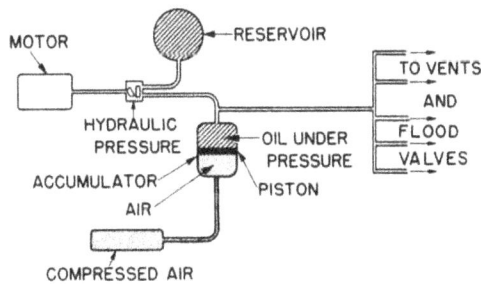

131.54

Figure 10-14.—This keeps pressure on tap.

CHAPTER 11

MACHINE ELEMENTS AND BASIC MECHANISMS

MACHINE ELEMENTS

Any machine, however simple, utilizes one or more basic machine elements or mechanisms in its makeup. In this chapter we will take a look at some of the more familiar elements and mechanisms used in naval machinery and equipment.

BEARINGS

In chapter 7 we saw that wherever two objects rub against each other, friction is produced. If the surfaces are very smooth, there will be little friction; if either or both are rough, there will be more friction. FRICTION is the resistance to any force that tends to produce motion of one surface over another. When you are trying to start a loaded hand truck rolling, you have to give it a hard tug (to overcome the resistance of static friction) to get it started. Starting to slide the same load across the deck would require a harder push than starting it on rollers. That is because rolling friction is always less than sliding friction. To take advantage of this fact, rollers or bearings are used in machines to reduce friction. Lubricants on bearing surfaces reduce the friction even further.

A bearing is a support and guide which carries a moving part (or parts) of a machine and maintains the proper relationship between the moving part or parts and the stationary part. It usually permits only one form of motion, as rotation, and prevents any other. There are two basic types of bearings: sliding type (plain bearings), also called friction or guide bearings, and antifrictional type (roller and ball bearings).

Sliding Type (Plain) Bearings

In bearings of this type a film of lubricant separates the moving part from the stationary part. There are three types of sliding motion bearings in common use: Reciprocal motion bearings, journal bearings, and thrust bearings.

1. RECIPROCAL MOTION BEARINGS provide a bearing surface on which an object slides back and forth. They are found on steam reciprocating pumps, where connecting rods slide on bearing surfaces near their connections to the pistons. Similar bearings are used on the connecting rods of large internal-combustion engines, and in many mechanisms operated by cams.

2. JOURNAL BEARINGS are used to guide and support revolving shafts. The shaft revolves in a housing fitted with a liner. The inside of the liner, on which the shaft bears, is made of babbitt metal or similar soft alloy (antifriction metal) to reduce friction. The soft metal is backed by a bronze or a copper layer, and that has a steel back for strength. Sometimes the bearing is made in two halves, and is clamped or screwed around the shaft (fig. 11-1). It is also called a laminated sleeve bearing.

Under favorable conditions the friction in journal bearings is remarkably small. However, when the rubbing speed of a journal bearing is very low or extremely high, the friction loss may become excessive when compared with the performance of a rolling surface bearing. A good example is the railroad car, now being fitted with roller bearings to eliminate the "hot box" troubles of journal bearings.

Heavy-duty bearings have oil circulated around and through them and some have an additional cooling system that circulates water around the bearing. Although revolving the steel shaft against babbitt metal produces less friction (and therefore less heat and wear) than steel against steel, it is still a problem to keep the parts cool. You know what causes a "burned out bearing" on your car, and how to prevent it. The same care and lubrication are necessary on all Navy equipment, only more so, because

69

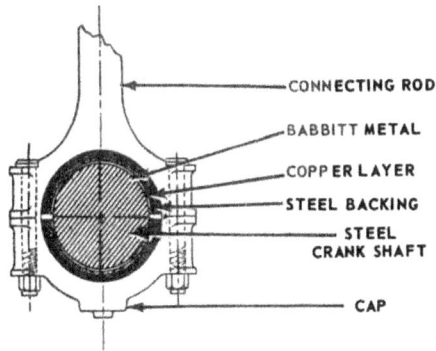

5.20

Figure 11-1.—Babbitt-lined bearing in which steel shaft revolves.

there is a lot of equipment, and many lives depend on its continued operation.

3. THRUST BEARINGS are used on rotating shafts, such as those supporting bevel gears, worm gears, propellers, and fans. They are installed to resist axial thrust or force and to limit axial movement. They are used chiefly on heavy machinery, such as Kingsbury thrust bearings used in heavy marine propelling machinery (figs. 11-2 and 11-3). The base of the housing holds an oil bath, and the rotation of the shaft continually distributes the oil. The bearing consists of a thrust collar on the propeller shaft and two or more stationary thrust shoes on either side of the collar. Thrust is transmitted from the collar through the shoes to the gear housing and the ship's structure to which the gear housing is bolted.

Antifrictional Or Roller
and Ball Bearings

You have had first-hand acquaintance with ball bearings since you were a child. They are what made your roller skates or bicycle wheels spin freely. If any of the little steel balls came out and were lost, your roller skates screeched and groaned. The balls or rollers are of hard, highly polished steel. The typical bearing consists of two hardened steel rings (called RACES), the hardened steel balls or rollers, and a a SEPARATOR. The motion occurs between the race surfaces and the rolling elements. There

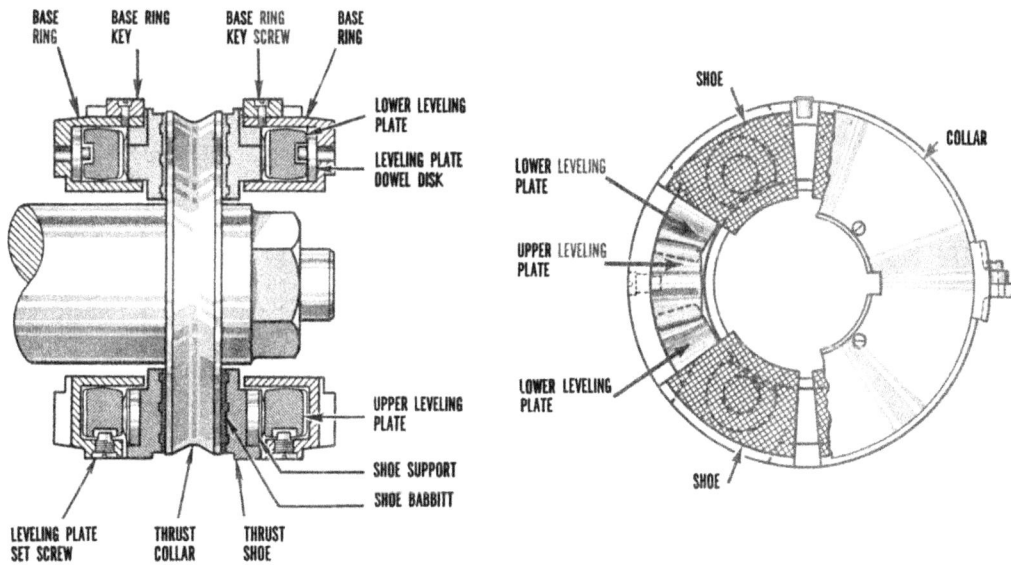

38.85:.86X

Figure 11-2.—Kingsburg pivoted-shoe thrust bearing.

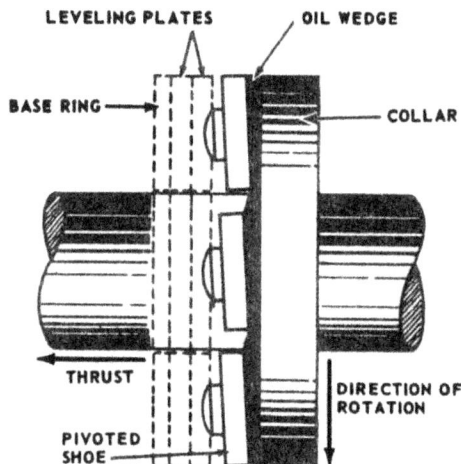

38.86.1

Figure 11-3.—Diagrammatic arrangement of a Kingsburg thrust bearing, showing oil film.

are seven basic types of antifrictional bearings (fig. 11-4).

1. Ball bearings
2. Cylindrical roller bearings
3. Tapered roller bearings
4. Self-aligning roller bearings with spherical outer raceway
5. Self-aligning roller bearings with spherical inner raceway
6. Ball thrust bearings
7. Needle roller bearings

Roller bearing assemblies are usually easy to disassemble for inspection, cleaning, and replacement of parts. Ball bearings, however, are assembled by the manufacturer and installed, or replaced, as a unit. Sometimes maintenance publications refer to roller and ball bearings as being either thrust or radial bearings. The difference between the two depends on the angle of intersection between the direction of the load and the plane of rotation of the bearing. Figure 11-5A shows a radial ball bearing assembly. The load here is pressing outward along the radius of the shaft. Now suppose a strong thrust were to be exerted on the right end of the shaft, tending to move it to the left. You can see that the radial bearing is not designed to support this axial thrust. Even putting a shoulder between the load and the inner race wouldn't do. It would just

pop the bearings out of their races. The answer is to arrange the races differently, as in figure 11-5B. Here is a thrust bearing. With a shoulder under the lower race, and another between the load and the upper race, it will handle any axial load up to its design limit. Sometimes bearings are designed to support both thrust and radial loads. This is the explanation of the term RADIAL THRUST bearings. The tapered roller bearing in figure 11-6 is an example.

Antifriction bearings require smaller housings than other bearings of the same load capacity, and can operate at higher speeds.

SPRINGS

Springs are elastic bodies (generally metal) which can be twisted, pulled, or stretched by some force, and which have the ability to return to their original shape when the force is released. All springs used in naval machinery are made of metal—usually steel, though some are of phosphor bronze, brass, or other alloys. A part that is subject to constant spring thrust or pressure is said to be SPRING LOADED. (Some components that appear to be spring loaded are actually under hydraulic or pneumatic pressure, or are moved by weights.)

Functions of Springs

Springs are used for many purposes, and one spring may serve more than one purpose. Listed below are some of the more common of these functional purposes. As you read them, try to think of at least one familiar application of each.

1. To store energy for part of a functioning cycle.
2. To force a component to bear against, to maintain contact with, to engage, to disengage, or to remain clear of, some other component.
3. To counterbalance a weight or thrust (gravitational, hydraulic, etc.). Such springs are usually called equilibrator springs.
4. To maintain electrical continuity.
5. To return a component to its original position after displacement.
6. To reduce shock or impact by gradually checking the motion of a moving weight.
7. To permit some freedom of movement between aligned components without disengaging them. These are sometimes called takeup springs.

71

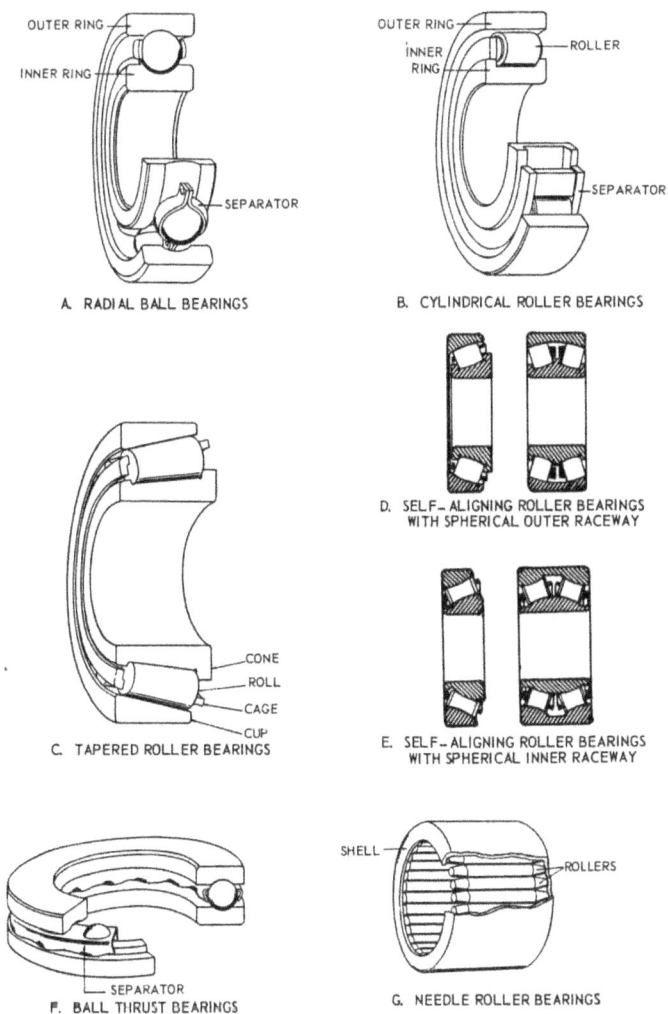

A. RADIAL BALL BEARINGS

B. CYLINDRICAL ROLLER BEARINGS

C. TAPERED ROLLER BEARINGS

D. SELF-ALIGNING ROLLER BEARINGS WITH SPHERICAL OUTER RACEWAY

E. SELF-ALIGNING ROLLER BEARINGS WITH SPHERICAL INNER RACEWAY

F. BALL THRUST BEARINGS

G. NEEDLE ROLLER BEARINGS

5.21

Figure 11-4.—The seven basic types of antifrictional bearings.

Types of Springs

As you read different books you will find that authors do not agree on classification of types of springs. The names are not as important as the types of work they can do and the loads they can bear. We may say there are three basic types: (1) flat; (2) spiral; (3) helical or coil.

1. FLAT springs include various forms of elliptic or leaf springs (fig. 11-7A (1&2)), made up of flat or slightly curved bars, plates, or leaves, and special flat springs (fig. 11-7A (3)). A special flat spring is made from a flat strip or bar, into whatever shape or design is calculated to be best suited for its position and purpose.

A

B

5.21

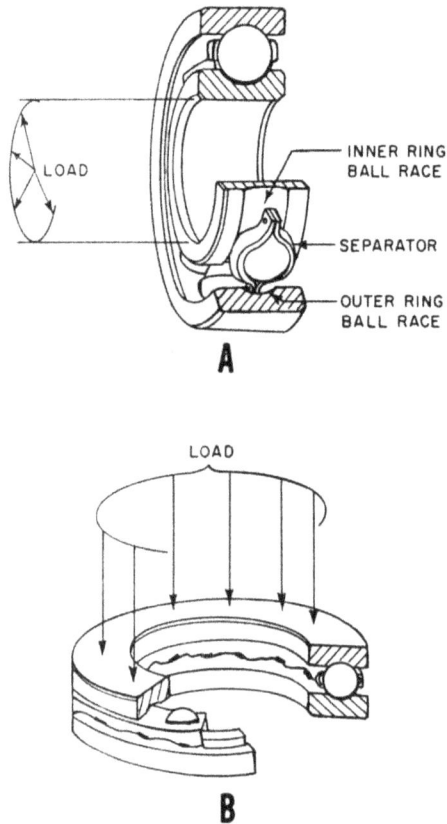

Figure 11-5.—Ball bearings. A. Radial type; B. Thrust type.

84.120

Figure 11-6.—Radial-thrust roller bearing.

2. SPIRAL springs are sometimes called clock or power springs (11-7B), and sometimes coil springs. A well known example is a watch or clock spring, which is wound (tightened) and then gradually releases the power as it unwinds.

Although there is good authority for calling this spring by other names, to avoid confusion we shall consistently call it SPIRAL.

3. HELICAL springs, often called spiral, but not in this text (fig. 11-7D), are probably the most common type of spring. They may be used in compression (fig. 11-7D (l)), extension or tension (fig. 11-7D (2), or torsion (fig. 11-7D (3)). A spring used in compression tends to shorten in action, while a tension spring

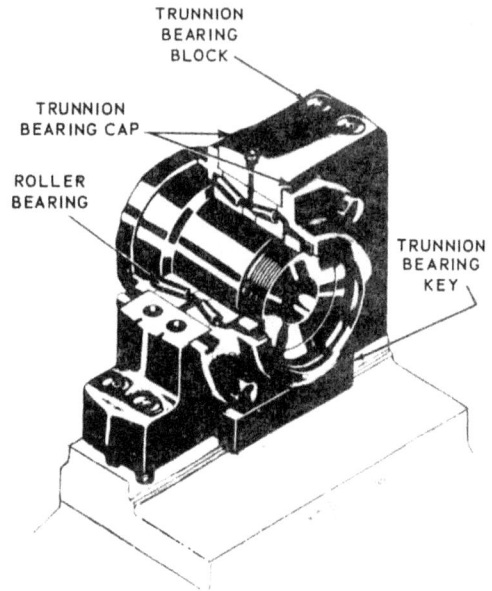

lengthens in action. Torsion springs are made to transmit a twist instead of a direct pull, and operate by coiling or uncoiling action.

In addition to straight helical springs, cone, double cone, keg, and volute springs are also classed as helical. These are usually used in compression. A cone spring (fig. 11-7D (4)), often called a valve spring because it is frequently used in valves, is shaped by winding the wire on a tapered mandrel instead of a straight one. A double cone spring (not illustrated) is composed of two cones joined at the small ends, and a keg spring (not illustrated) is two cone springs joined at their large ends.

VOLUTE springs (fig. 11-7D (5)) are conical springs made from a flat bar which is so wound that each coil partially overlaps the adjacent one. The width (and thickness) of the material gives it great strength or resistance.

A conical spring can be pressed flat so it requires little space, and it is not likely to buckle sidewise.

4. TORSION BARS (fig. 11-7C) are straight bars that are acted on by torsion (twisting force). The bar may be circular or rectangular in cross

73

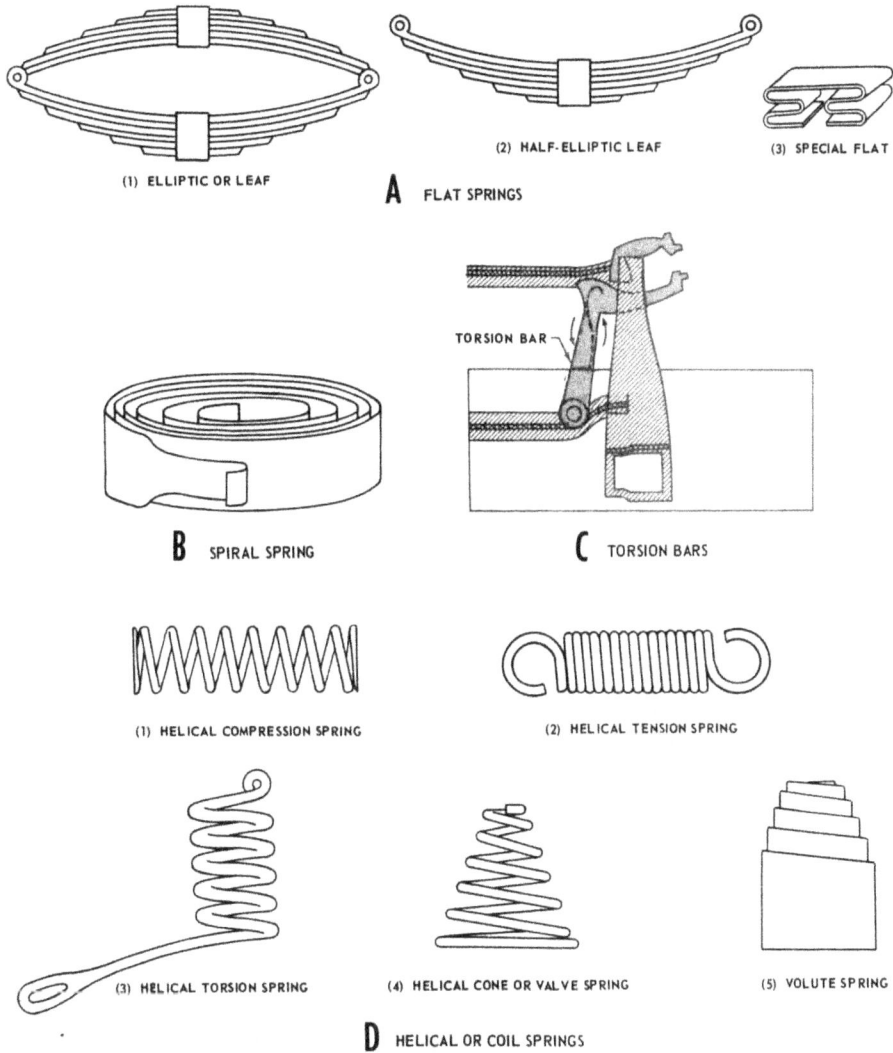

(1) ELLIPTIC OR LEAF
(2) HALF-ELLIPTIC LEAF
(3) SPECIAL FLAT

A FLAT SPRINGS

TORSION BAR

B SPIRAL SPRING

C TORSION BARS

(1) HELICAL COMPRESSION SPRING

(2) HELICAL TENSION SPRING

(3) HELICAL TORSION SPRING

(4) HELICAL CONE OR VALVE SPRING

(5) VOLUTE SPRING

D HELICAL OR COIL SPRINGS

84.131

Figure 11-7.—Types of springs.

section, or less commonly in other shapes. It may also be a tube.

5. A special type of spring is a RING SPRING or DISC spring (not illustrated). It is made of a number of metal rings or discs that overlap each other.

BASIC MECHANISMS

THE GEAR DIFFERENTIAL

A gear differential is a mechanism that is capable of adding and subtracting mechanically. To be more precise, it adds the total revolutions of two shafts—or subtracts the total revolutions of one shaft from the total revolutions of another shaft—and delivers the answer by positioning a third shaft. The gear differential will add or subtract any number of revolutions, or very small fractions of revolutions, continuously and accurately. It will produce a continuous series of answers as the inputs change.

Figure 11-8 is a cutaway drawing of a bevel gear differential showing all its parts and how they are related to each other. Grouped around the center of the mechanism are four bevel gears, meshed together. The two bevel gears on either side are called "end gears." The two bevel gears above and below are called "spider gears." The long shaft running through the end gears and the three spur gears is called the "spider shaft." The short shaft running through the spider gears, together with the spider gears themselves, is called the "spider."

Each of the spider gears and the end gears are bearing mounted on their shafts and are free to rotate. The spider shaft is rigidly connected with the spider cross shaft at the center block where they intersect. The ends of the spider shaft are secured in flanges or hangers, but they are bearing mounted and the shaft is free to rotate on its axis. It follows then that to rotate the spider shaft, the spider, consisting of the spider cross shaft and the spider gears, must tumble, or spin, on the axis of the spider shaft, inasmuch as the two shafts are rigidly connected.

The three spur gears shown in figure 11-8 are used to connect the two end gears and the spider shaft to other mechanisms. They may be of any convenient size. Each of the two input spur gears is attached to an end gear. An input gear and an end gear together are called a "side" of a differential. The third spur gear is the output gear, as designated in figure 11-8. This is the

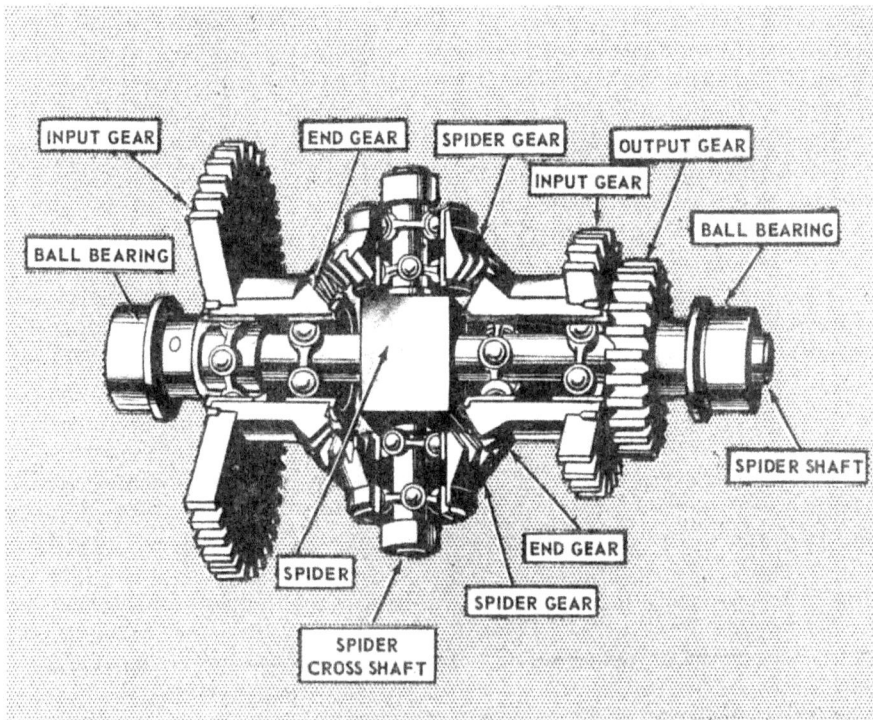

12.87

Figure 11-8.—Bevel gear differential.

75

only gear that is pinned to the spider shaft. All of the other gears, both bevel and spur, in the differential are bearing mounted.

Figure 11-9 is an exploded view of a gear differential showing each of its individual parts, and figure 11-10 is a schematic sketch showing the relationship of the principle parts.

How it Works

For the present we will assume that the two sides are the inputs and the gear on the spider shaft is the output. Later it will be shown that any of these three gears can be either an input or an output. Now let's look at figure 11-11. In this hookup the two end gears are positioned by the input shafts, which represent the quantities to be added or subtracted. The spider gears do the actual adding and subtracting. They follow the rotation of the two end gears, turning the spider shaft a number of revolutions proportional to the sum, or difference, of the revolutions of the end gears.

Suppose the left side of the differential is rotated while the other remains stationary, as in block 2 of figure 11-11. The moving end gear will drive the spider gears, making them roll on the stationary right end gear. This motion will turn the spider in the same direction as the input and, through the spider shaft and output gear, the output shaft. The output shaft will turn a number of revolutions proportional to the input.

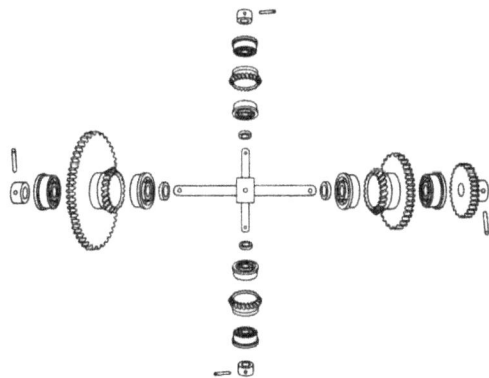

110.8

Figure 11-10.—The differential. End gears and spider arrangement.

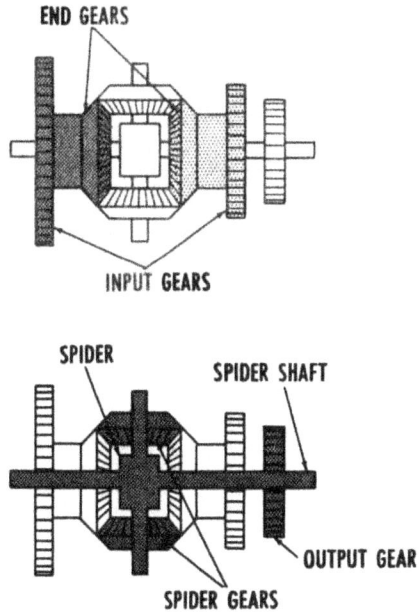

131.55

Figure 11-9.—Exploded view of differential gear system.

If the right side is now rotated and the left side held stationary, as in block 3 of figure 11-11, the same thing will happen. If both input sides of the differential are turned in the same direction at the same time, the spider will be turned by both at once, as in block 4 of figure 11-11. The output will be proportional to the sum of the two inputs. Actually, the spider makes only half as many revolutions as the sum of the revolutions of the end gears, because the spider gears are free to roll between the end gears. To understand this better, let's look at figure 11-12. Here a cylindrical drinking glass is rolled along a table top by pushing a ruler across its upper side. The glass will roll only half as far as the ruler travels. The spider gears in the differential roll against the end gears in exactly the same way. Of course, the answer can be corrected by

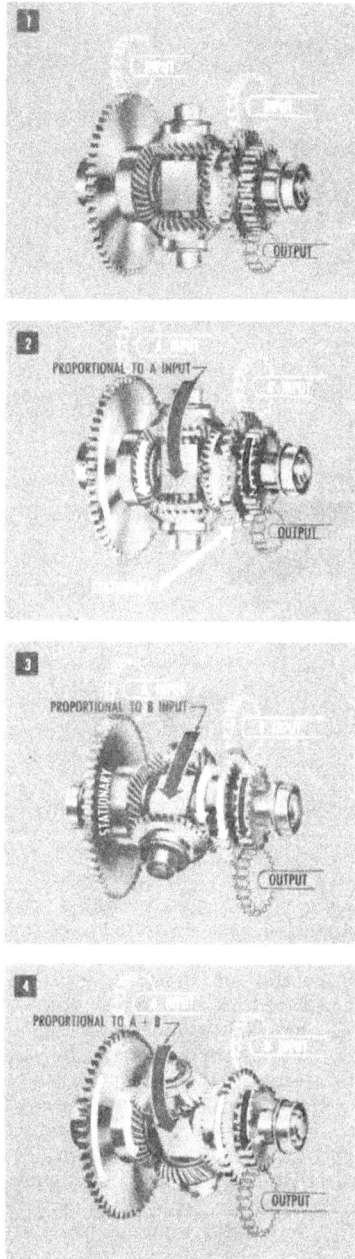

131.56

Figure 11-12.—The spider makes only half as many revolutions.

using a 2:1 gear ratio between the gear on the spider shaft and the gear for the output shaft. Very often, for design purposes, this gear ratio will be found to be different.

When the two sides of the differential move in opposite directions, the output of the spider shaft is proportional to the difference of the revolutions of the two inputs. This is because the spider gears are free to turn, and are driven in opposite directions by the two inputs. If the two inputs are equal and opposite, the spider gears will turn, but there will be no movement of the spider shaft. If the two inputs turn in opposite directions for an unequal number of revolutions, the spider gears roll on the end gear that makes the lesser number of revolutions, rotating the spider in the direction of the input making the greater number of revolutions. The motion of the spider shaft will be equal to half the difference between the revolutions of the two inputs. A change in the gear ratio to the output shaft can then give us any proportional answer we wish.

We have thus far been describing a hookup wherein the two sides are inputs and the spider shaft the output. As long as it is recognized that the spider follows the end gears for half the sum, or difference, of their revolutions, however, it is not necessary to always use this type hookup. The spider shaft may be used as one input and either of the sides used as the other. The other side will then become the output. This fact permits three different hookups for any given differential, as is illustrated in figure 11-13. Whichever proves the most convenient mechanically may be used.

In chapter 14 of this book, the differential as used in the automobile will be described. This differential is similar in principle, but, as you

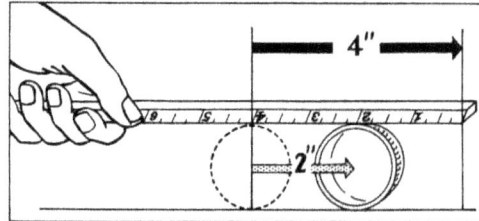

110.9

Figure 11-11.—How a differential works.

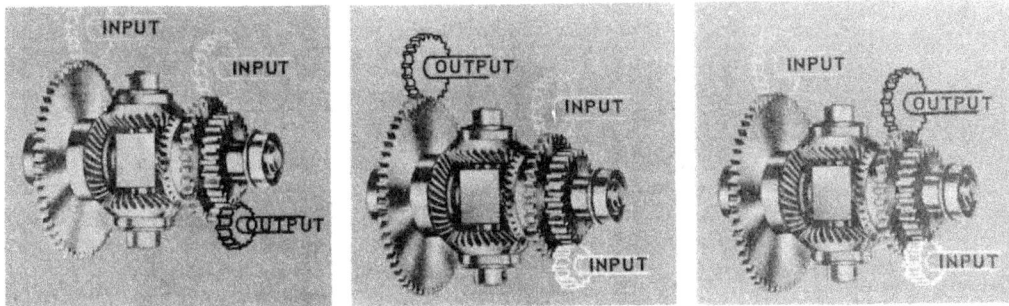

110.9

Figure 11-13.—Any of these three hookups can be used.

will see, is somewhat different in its mechanical makeup. In chapter 15 you will be given information on differentials as they are used in computers.

LINKAGES

A linkage may consist of either one or a combination of the following basic parts:
1. Rod, shaft, or plunger
2. Lever
3. Rocker arm
4. Bell crank

These parts combined are used to transmit limited rotary or linear motion. To change the direction of a motion, cams are used with the linkage.

Lever type linkages (fig. 11-14) are used in equipment which has to be opened and closed; for instance, valves in electric-hydraulic systems, gates, clutches, clutch-solenoid interlocks, etc. Rocker arms are merely a variation, or special use, of levers.

Bell cranks are used primarily to transmit motion from a link traveling in one direction to another link which is to be moved in a different direction. The bell crank is mounted on a fixed pivot, and the two links are connected at two points in different directions from the pivot. By properly locating the connection points, the output links can be made to move in any desired direction.

All linkages require occasional adjustments or repair, particularly when they become worn. To make the proper adjustments, a person must be familiar with the basic parts which constitute a linkage. Adjustments are normally made by lengthening or shortening the rods and shafts by means of a clevis or turnbuckle.

COUPLINGS

In a broad sense, the term "coupling" applies to any device that holds two parts together. Line shafts which are made up of several shafts of different lengths may be held together by any of several types of shaft couplings. When shafts are very closely aligned, the sleeve coupling, as in figure 11-15, may be used. It consists of a metal tube slit at each end. The slitted ends enable the clamps to fasten the sleeve securely to the shaft ends. With the clamps tightened, the shafts are held firmly together and turn as one shaft. The sleeve coupling also serves as a convenient device for making adjustments between units. The weight at the opposite end of the clamp from the screw is merely to offset the weight of the screw and clamp arms. By distributing the weight more evenly, shaft vibration is reduced.

The Oldham coupling, named for its inventor, may be used to transmit rotary motion between shafts which are parallel but not necessarily always in perfect alignment.

An Oldham coupling (fig. 11-16), consists of a pair of disks, one flat and the other hollow. These disks are pinned to the ends of the shafts. A third (center) disk, with a pair of lugs projecting from each face of the disk, fits into the slots between the two end disks and thus enables one shaft to drive the other shaft. A coil spring, housed within the center and the hollow end disk,

78

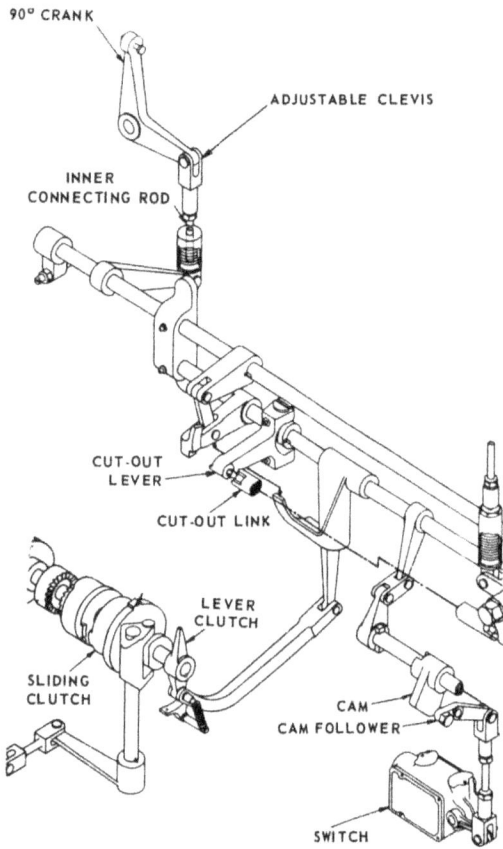

131.57

Figure 11-14.—Linkages.

forces the center disk against the flat disk. When the coupling is assembled on the shaft ends, a flat lock spring is slipped into the space around the coil spring. The ends of the flat spring are formed so that when the flat spring is pushed into the proper place, the ends of the spring are pushed out and locked around the lugs. A lock wire is passed between the holes drilled through the projecting lugs to guard the assembly. The coil spring compensates for any change in shaft length. (Shaft length may vary due to changes in temperature.)

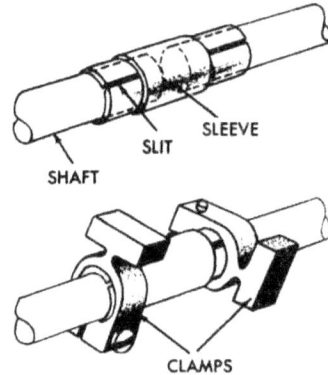

12.51

Figure 11-15.—Sleeve coupling.

The disks, or rings, connecting the shafts allow a small amount of radial play, and this allows a small amount of misalignment of the shafts as they rotate. Oldham type couplings can be easily connected and disconnected.

A universal joint is the answer when two shafts not in the same plane must be coupled. Universal joints may have various forms. They are used in nearly all types and classes of machinery. An elementary universal joint, sometimes called a Hooke joint (fig. 11-17), consists of two U-shaped yokes fastened to the ends of the shafts to be connected. Within these yokes is a cross-shaped part which holds the yokes together and allows each yoke to bend, or pivot, one with respect to the other. With this arrangement, one shaft can drive the other even though the angle between the two is as great as 25° from alignment. Figure 11-18 shows a ring and trunnion type of universal joint. This is merely a slight modification of the old Hooke joint. This type is commonly used in automobile drive shaft systems. Two, and sometimes three, are utilized. You will read more about these in chapter 14 of this book. Another type of universal joint is used where a smoother torque transmission is desired and less structural strength is required. This is the Bendix-Weiss universal joint (fig. 11-19). In this type of joint, four large balls transmit the rotary force, with a smaller ball as a spacer. With the Hooke type of universal joint, a whipping motion occurs as the shafts rotate—the amount of whip depending on the degree of shaft misalignment. The Bendix-Weiss joint does not have this disadvantage; it

79

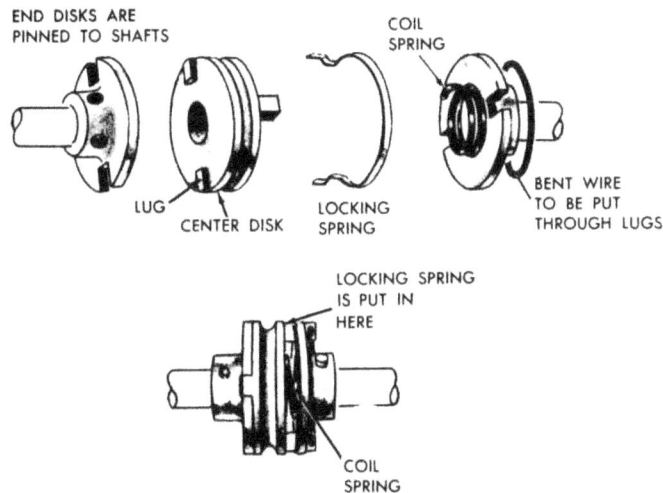

Figure 11-16.—Oldham coupling.

12.52

transmits rotary motion with a constant angular velocity. This type of joint is both more expensive to manufacture and of less strength than the Hooke types, however.

The following four types of couplings are also used extensively in naval equipment:

1. The fixed (sliding lug) coupling is nonadjustable; however, it does allow for a small amount of misalignment in shafting (fig. 11-20).

2. The flexible coupling (fig. 11-21), connects two shafts by means of a metal disk. Two coupling hubs, each splined to its respective shaft, are bolted to the metal disk. The flexible coupling provides a small amount of flexibility to allow for a slight axial misalignment of the shafts.

3. The adjustable (vernier) coupling provides a means of finely adjusting the relationship of two interconnected rotating shafts, (fig. 11-22). By loosening a clamping bolt and turning an adjusting worm, one shaft may be rotated while the other remains stationary. When the proper relationship is attained, the clamping bolt is retightened, locking the shafts together again.

4. The adjustable flexible (vernier) coupling (fig. 11-23) is simply a combination of the flexible disk coupling and the adjustable (vernier) coupling.

CAM AND CAM FOLLOWERS

A cam is a specially shaped surface, projection, or groove whose movement with respect to a part in contact with it (cam follower) drives the cam follower in another movement in response. A cam may be a projection on a revolving shaft (or on a wheel) for the purpose of changing the direction of motion from rotary to up-and-down, or vice versa. It may be a sliding piece or a groove to impart an eccentric motion. Some cams do not move at all, but cause a change of motion in the contacting part. Cams are not ordinarily used to transmit power in the sense that gear trains are. They are generally used to modify mechanical movement, the power for which is furnished through other means. They may control other mechanical units, or lock together or synchronize two or more engaging units.

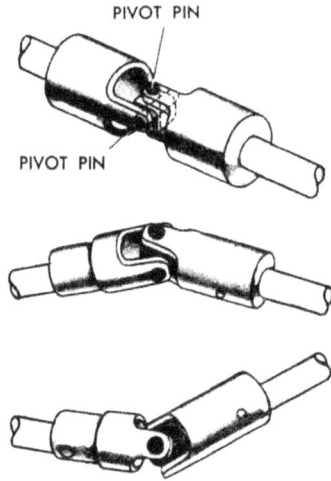

5.34

Figure 11-17.—Universal joint (Hooke type).

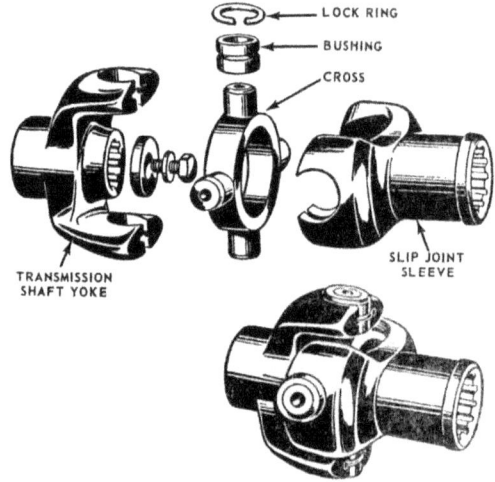

81.194

Figure 11-18.—Ring and trunnion universal joint.

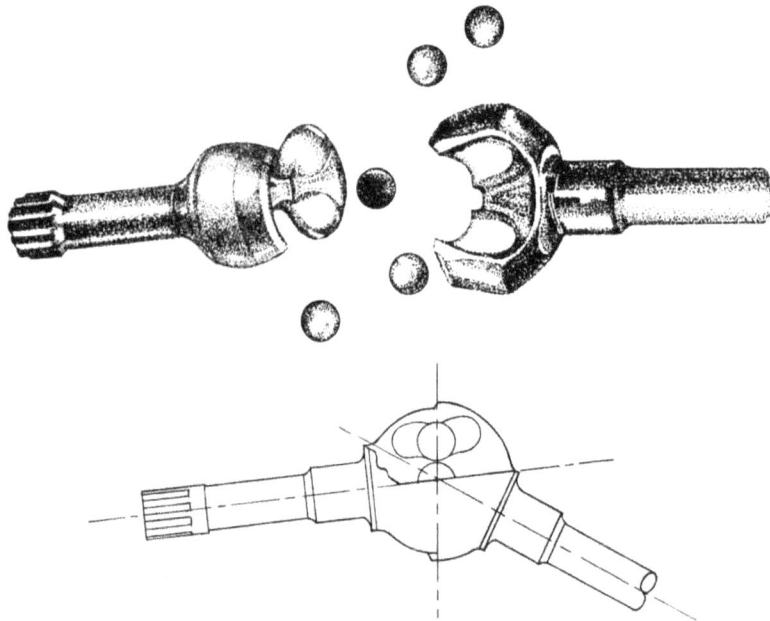

5.34

Figure 11-19.—Bendix-Weiss universal joint.

Edge or peripheral cams, also called disc cams, operate a mechanism in one direction only, gravity or a spring being relied upon to hold the cam roll in contact with the edge of the cam. The shape of the cam may be made to suit the action required, such as heart shape.

Face cams have a groove or roll path cut in the face and operate a lever or other mechanism positively in both directions, as the roll is always guided by the sides of the slot. Such a groove can be seen on top of the bolt of the Browning machine gun, caliber .30, or in fire control cams. The shape of the groove may give its name to the cam, as for example, constant lead cam, square cam, run-out cam.

The toe and wiper cam shown in figure 11-24 (d) is an example of a pivoted beam.

CLUTCHES

TYPES

A clutch is a form of coupling which is designed to connect or disconnect a driving and a driven member for stopping or starting the driven part. There are two general classes of clutches—positive clutches and friction clutches.

Positive Clutches. Positive clutches have teeth which interlock. The simplest is the jaw or claw type (fig. 11-25A), which is usable only at low speeds. The spiral claw or ratchet type (fig. 11-25B) cannot be reversed. An example of a clutch is seen in bicycles—it engages the rear sprocket with the rear wheel when the pedals are pushed forward, and lets the rear wheel revolve freely when the pedals are stopped.

Friction Clutches. The object of a friction clutch is to connect a rotating member to one that is stationary, to bring it up to speed, and to transmit power with a minimum of slippage. Figure 11-25C shows a cone clutch commonly used in motor trucks. They may be single-cone or double-cone. Figure 11-25D shows a disc clutch, also used in autos. A disc clutch may also have a number of plates (multiple-disc clutch). In a series of discs, each driven disc is located between two driving discs. You may have had experience with a multiple-disc clutch on your car. The Hele-Shaw clutch is a combined conical-disc clutch (fig. 11-25E). The groove permits circulation of oil, and cooling. Single-disc clutches are frequently dry clutches (no lubrication); multiple-disc clutches may be dry or wet (lubricated or run in oil).

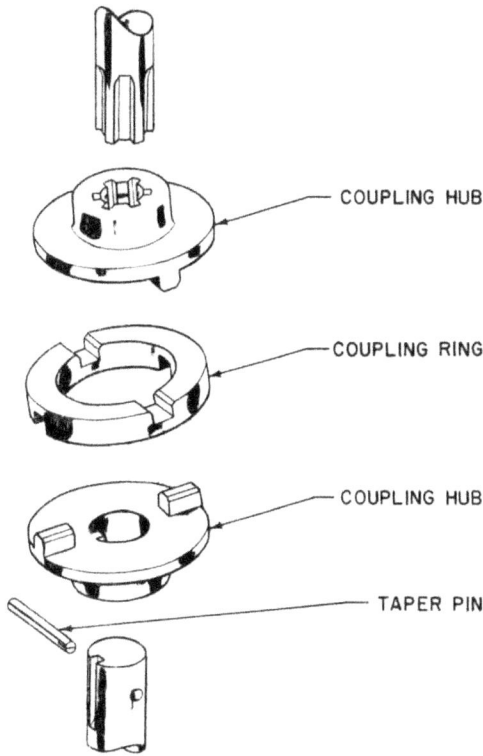

84.133.3
Figure 11-20.—Fixed coupling.

Types and Uses

Cams are of many shapes and sizes and are widely used in machines and machine tools (fig. 11-24). Cams may be classified as:
1. Radial or plate cams
2. Cylindrical or barrel cams
3. Pivoted beams

A similar grouping of types of cams is: Drum or barrel cams; edge cams; face cams.

The drum or barrel cam has a path for the roll or follower cut around the outside, and imparts a to-and-fro motion to a slide or lever in a plane parallel to the axis of the cam. Sometimes these cams are built up on a plain drum with cam plates attached.

Plate cams are used in 5''/38 and 3''/50 guns to open the breechblock during counter-recoil.

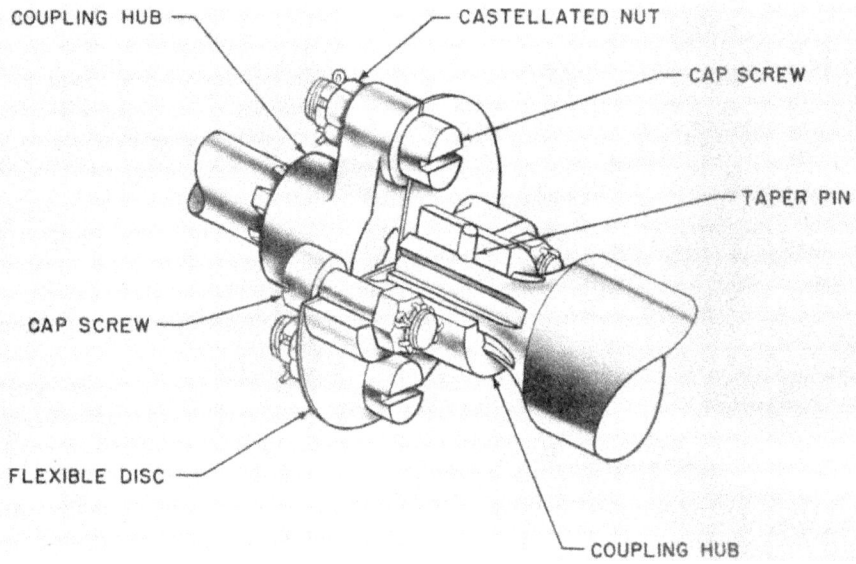

84.133.2

Figure 11-21.—Flexible coupling.

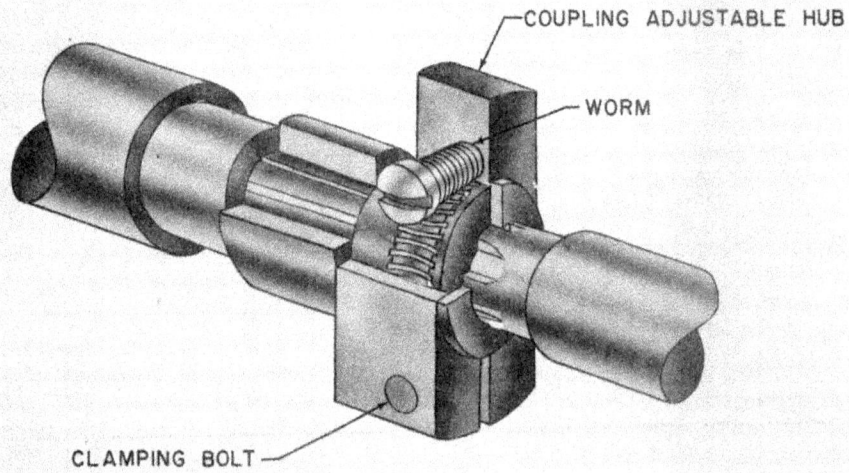

131.58

Figure 11-32.—Adjustable (vernier) coupling.

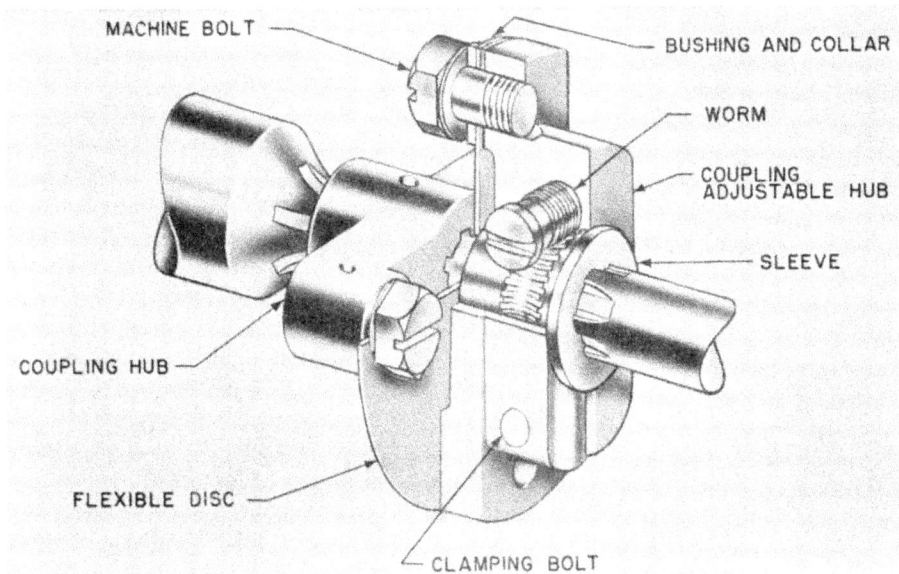

11.133.4

Figure 11-23.—Adjustable flexible (vernier) coupling.

Magnetic clutches are a recent development in which the friction surfaces are brought together by magnetic force when the electricity is turned on (fig. 11-25F). The induction clutch transmits power without contact between driving and driven members.

Expanding clutches or rim clutches are named according to the way the pressure is applied to the rim—block, split-ring, band, or roller. In one type of expanding clutch a powerful effect is gained by the expanding action of right-and left-hand screws as a sliding sleeve is moved along a shaft, and expands the band against the rim. The centrifugal clutch is a special application of a block clutch.

Coil clutches are used where heavy parts are to be moved, as in a rolling mill. Great friction is caused by the grip of the coil when it is thrust onto a cone on the driving shaft, yet the clutch is very sensitive to control.

Pneumatic and hydraulic clutches are used on Diesel engines and transportation equipment. Hydraulic couplings (fig. 11-25G), which serve also as clutches, are used in the hydraulic A-end of electric-hydraulic gun drives.

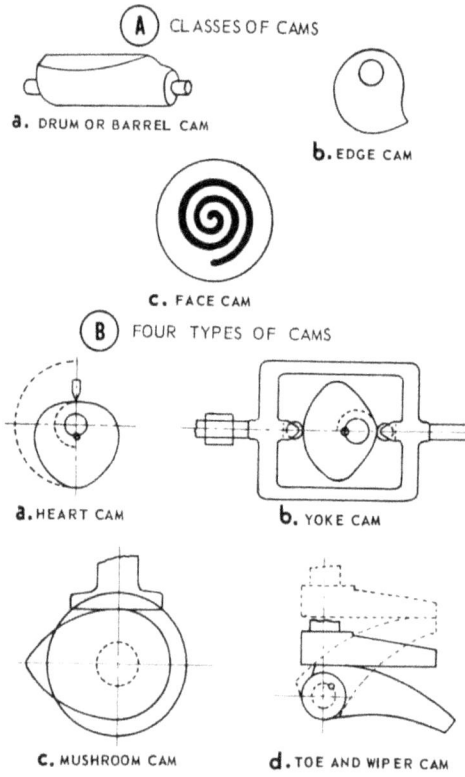

5.29

Figure 11-24.—Classes and types of cams.

(A) JAW OR CLAW TYPE

(B) SPIRAL CLAW OR RATCHET TYPE

FABRIC

FLYWHEEL

(C) CONE CLUTCH

(D) DISC CLUTCH

(E) COMBINED CONICAL–DISK CLUTCH

SWITCH CLOSED

SPRING PLATE

DRIVING SHAFT RUNNING

DRIVEN SHAFT RUNNING

COIL

SWITCH OPEN

FABRIC LINING

DRIVING SHAFT RUNNING

DRIVEN SHAFT STILL

COIL

(F) MAGNETIC CLUTCH

PRIMARY ROTOR (IMPELLER)

SECONDARY ROTOR (RUNNING)

EMPTYING HOLES

COVER OR ROTOR HOUSING

RING VALVE

(G) HYDRAULIC COUPLING

5.33

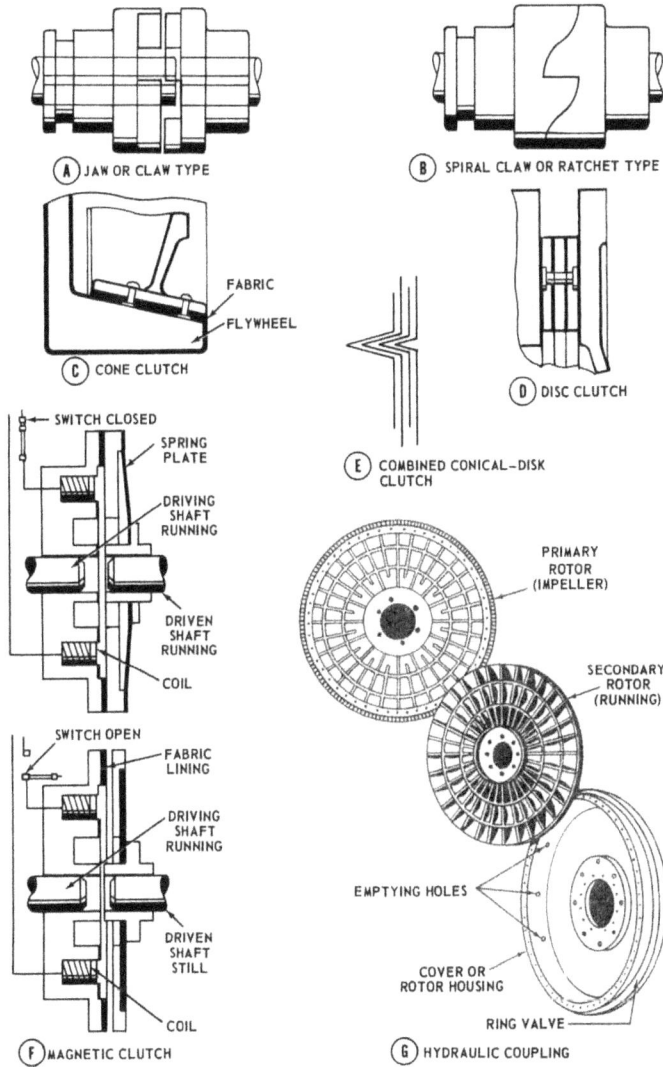

Figure 11-25.—Types of clutches.

Chapter 12, pages 88 through 105, have been deleted due to obsolete material.

CHAPTER 13

INTERNAL COMBUSTION ENGINE

The automobile is a familiar object to all of us; and the engine that makes it go is one of the most fascinating and talked about of all the complex machines we use today. In this chapter we will explain briefly some of the operational principles of this machine, and then break it down to its more basic mechanisms. In its makeup you will find many of the devices and basic mechanisms that you have studied earlier in this book. Look for these and the simple machines that make up the engine as you study its operation and construction.

COMBUSTION ENGINE

An engine is defined simply as a machine that converts heat energy to mechanical energy. To fulfill this purpose, the engine may take one of several forms.

Combustion is the act of burning. Internal means inside or enclosed. Thus an internal combustion engine is one in which the fuel burns inside; that is, burning takes place within the same cylinder that produces energy to turn the crankshaft. In external combustion engines, such as steam engines, the combustion takes place outside the engine. Figure 13-1 shows, in simplified form, an external and an internal combustion engine.

The external combustion engine requires a boiler to which heat is applied. This combustion causes water to boil to produce steam. The steam passes into the engine cylinder under pressure and forces the piston to move downward. With the internal combustion engine, the combustion takes place inside the cylinder and is directly responsible for forcing the piston to move downward.

The transformation of heat energy to mechanical energy by the engine is based on a fundamental law of physics which states that gas will expand upon application of heat. The law also states that when a gas is compressed the temperature of the gas will increase. If the gas is confined with no outlet for expansion, then the pressure of the gas will be increased when heat is applied (as it is in an automotive cylinder). In an engine, this pressure acts against the head of a piston, causing it to move downward.

As you know, the piston moves up and down in the cylinder. The up-and-down motion is known as reciprocating motion. This reciprocating motion (straight line motion) must be changed to rotary motion (turning motion) in order to turn the wheels of a vehicle. A crank and a connecting rod change this reciprocating motion to rotary motion.

All internal combustion engines, whether gasoline or diesel, are basically the same. We can best demonstrate this by saying they all rely on three things—air, fuel, and ignition.

Fuel contains potential energy for operating the engine; air contains the oxygen necessary for combustion; and ignition starts combustion. All are fundamental, and the engine will not operate without any one of them. Any discussion of engines must be based on these three factors and the steps and mechanisms involved in delivering them to the combustion chamber at the proper time.

DEVELOPMENT OF POWER

The power of an internal combustion engine comes from the burning of a mixture of fuel and air in a small, enclosed space. When this mixture burns it expands greatly, and the push or pressure created is used to move the piston, thereby cranking the engine. This movement is eventually sent back to the wheels to drive the vehicle.

Since similar action occurs in all cylinders of an engine, let's use one cylinder in our development of power. The one-cylinder engine

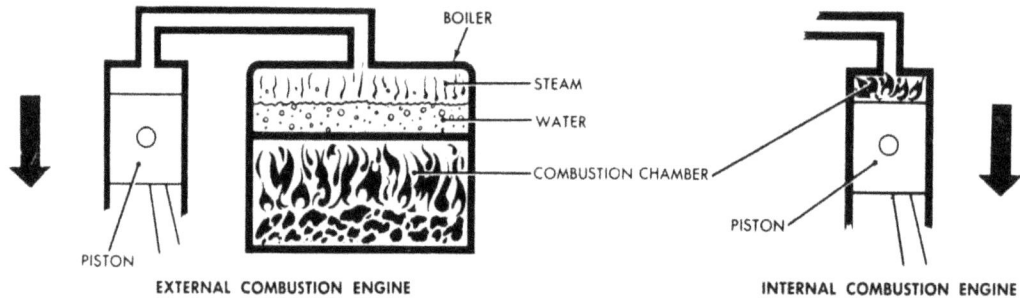

Figure 13-1.—Simple external and internal combustion engine.

81.41

consists of four basic parts as shown in figure 13-2.

First we must have a cylinder which is closed at one end; this cylinder is similar to a tall metal can.

Inside the cylinder is the piston, a movable metal plug, which fits snugly into the cylinder, but can still slide up and down easily. This up-and-down movement, produced by the burning of fuel in the cylinder, results in the production of power from the engine.

You have already learned that the up-and-down movement is called reciprocating motion. This motion must be changed to rotary motion

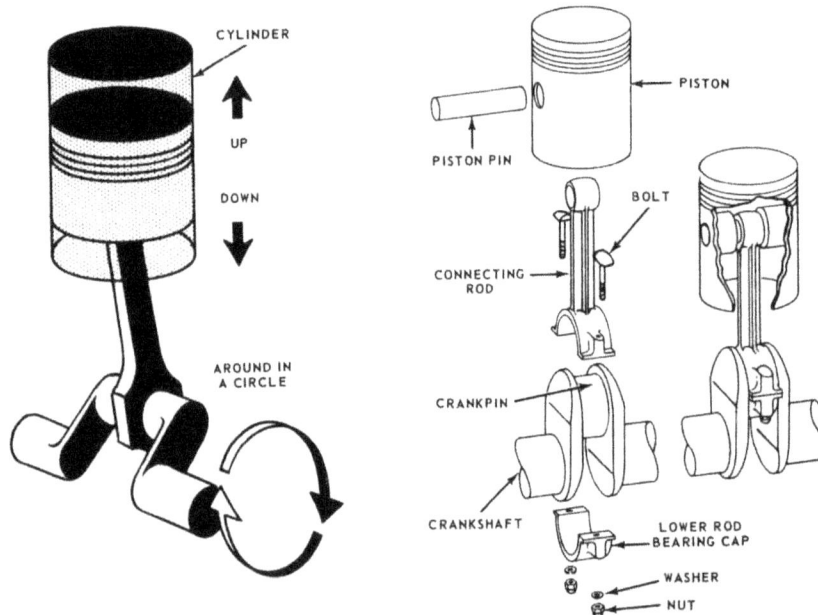

65.95

Figure 13-2.—Cylinder, piston, connecting rod, and crankshaft for a one-cylinder engine.

107

so the wheels or tracks of vehicles can be made to rotate. This change is accomplished by a crank on the crankshaft and a connecting rod which connects between the piston and the crank.

The crankshaft is a shaft with an offset portion, the crank, which describes a circle as the shaft rotates. The top end of the connecting rod is connected to the piston and must therefore go up and down. The lower end of the connecting rod is attached to the crankshaft. The lower end of the connecting rod also moves up and down but, because it is attached to the crankshaft, it must also move in a circle with the crank.

When the piston of the engine slides downward because of the pressure of the expanding gases in the cylinder, the upper end of the connecting rod moves downward with the piston, in a straight line. The lower end of the connecting rod moves down and in a circular motion at the same time. This moves the crank and in turn the crank rotates the shaft; this rotation is the desired result. So remember, the crankshaft and connecting rod combination is a mechanism for the purpose of changing straightline, up-and-down motion to circular, or rotary motion.

BASIC ENGINE STROKES

Each movement of the piston from top to bottom or from bottom to top is called a stroke. The piston takes two strokes (an upstroke and a downstroke) as the crankshaft makes one complete revolution. When the piston is at the top of a stroke, it is said to be at top dead center (TDC). When the piston is at the bottom of a stroke, it is said to be at bottom dead center (BDC). These positions are called rock positions and will be discussed further in this chapter under "Timing." See figure 13-3 and figure 13-7.

The basic engine you have studied so far has had no provisions for getting the fuel-air mixture into the cylinder or burned gases out of the cylinder. There are two openings in the enclosed end of a cylinder. One of the openings, or ports, permits the mixture of air and fuel to enter and the other port permits the burned gases to escape from the cylinder. The two ports have valves assembled in them. These valves, actuated by the camshaft, close off either one or the other of the ports, or both of them, during various stages of engine operation. One of the valves, called the intake valve, opens to admit a mixture of fuel and air into the cylinder.

The other valve, called the exhaust valve, opens to allow the escape of burned gases after the fuel-and-air mixture has burned. Later on you will learn more about how these valves and their mechanisms operate.

The following paragraphs give a simplified explanation of the action that takes place within the engine cylinder. This action may be divided into four parts: the intake stroke, the compression stroke, the power stroke, and the exhaust stroke. Since these strokes are easy to identify in the operation of a four-cycle engine, that engine is used in the description. This type of engine is also called a four-stroke-Otto-cycle engine, because it was Dr. N. A. Otto who, in 1876, first applied the principle of this engine.

INTAKE STROKE

The first stroke in the sequence is called the intake stroke (fig. 13-4). During this stroke, the piston is moving downward and the intake valve is open. This downward movement of the piston produces a partial vacuum in the cylinder, and air and fuel rush into the cylinder past the open intake valve. This is somewhat the same effect as when you drink through a straw. A partial vacuum is produced in the mouth and the liquid moves up through the straw to fill the vacuum.

COMPRESSION STROKE

When the piston reaches bottom dead center at the end of the intake stroke and is therefore at the bottom of the cylinder, the intake valve closes. This seals the upper end of the cylinder. As the crankshaft continues to rotate, it pushes up, through the connecting rod, on the piston. The piston is therefore pushed upward and compresses the combustible mixture in the cylinder; this is called the compression stroke (fig. 13-4). In gasoline engines, the mixture is compressed to about one-eighth of its original volume. (In a diesel engine the mixture may be compressed to as little as one-sixteenth of its original volume.) This compression of the air-fuel mixture increases the pressure within the cylinder. Compressing the mixture in this way makes it still more combustible; not only does the pressure in the cylinder go up, but the temperature of the mixture also increases.

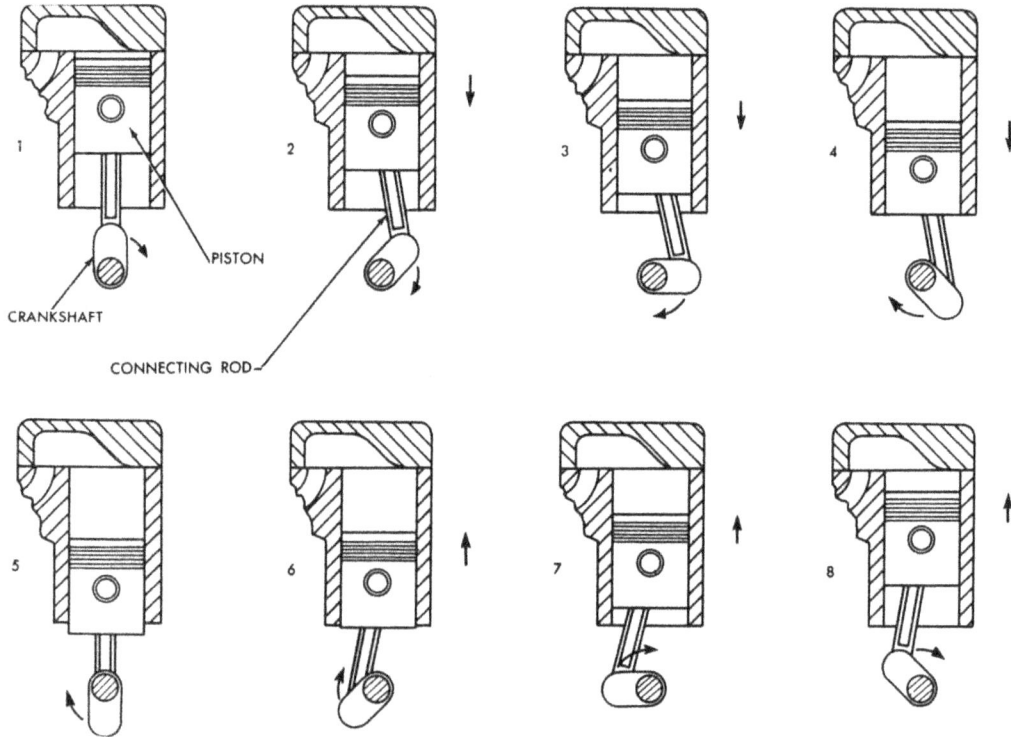

81.42

Figure 13-3.—Relationship of piston, connecting rod, and crank on crankshaft as crankshaft turns one revolution.

POWER STROKE

As the piston reaches top dead center at the end of the compression stroke and therefore has moved to the top of the cylinder, the compressed fuel-air mixture is ignited. The ignition system causes an electric spark to occur suddenly in the cylinder, and the spark sets fire to the fuel-air mixture. In burning, the mixture gets very hot and tries to expand in all directions. The pressure rises to about 600 or 700 pounds per square inch. Since the piston is the only thing that can move, the force produced by the expanding gases forces the piston down. This force, or thrust, is carried through the connecting rod to the crankpin on the crankshaft. The crankshaft is given a powerful twist. This is called the power stroke (fig. 13-4). This turning effort, rapidly repeated in the engine and carried through gears and shafts, will turn

the wheels of a vehicle and cause it to move along the highway.

EXHAUST STROKE

After the fuel-air mixture has burned, it must be cleared from the cylinder. This is done by opening the exhaust valve just as the power stroke is finished and the piston starts back up on the exhaust stroke (fig. 13-4). The piston forces the burned gases out of the cylinder past the open exhaust valve. The four strokes (intake, compression, power, and exhaust) are continuously repeated as the engine runs.

ENGINE CYCLES

Now, with the basic knowledge you have of the parts and the four strokes of the engine, let

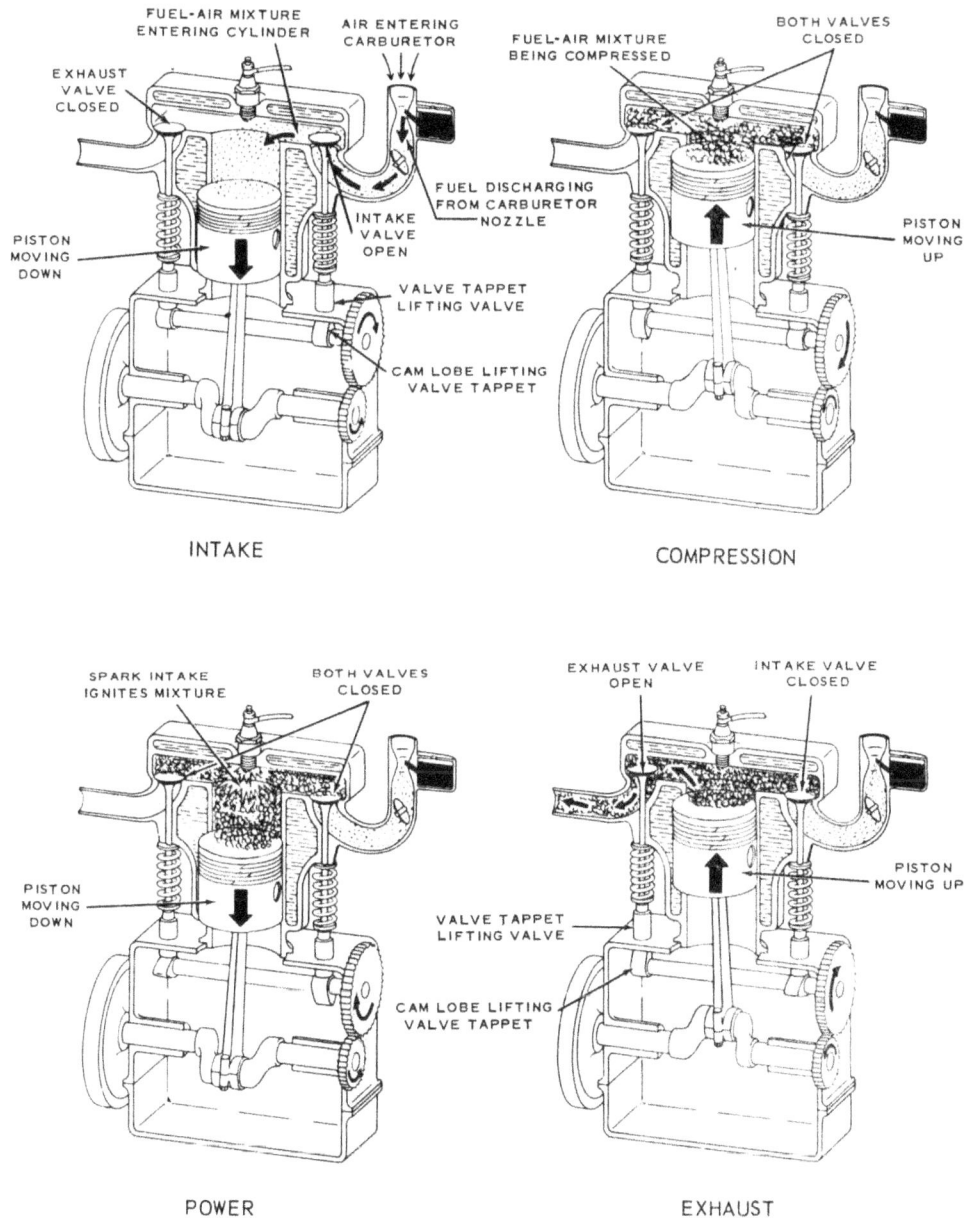

FUEL-AIR MIXTURE
ENTERING CYLINDER

AIR ENTERING
CARBURETOR

EXHAUST
VALVE
CLOSED

INTAKE
VALVE
OPEN

PISTON
MOVING
DOWN

FUEL DISCHARGING
FROM CARBURETOR
NOZZLE

VALVE TAPPET
LIFTING VALVE

CAM LOBE LIFTING
VALVE TAPPET

INTAKE

BOTH VALVES
CLOSED

FUEL-AIR MIXTURE
BEING COMPRESSED

PISTON
MOVING
UP

COMPRESSION

SPARK INTAKE
IGNITES MIXTURE

BOTH VALVES
CLOSED

PISTON
MOVING
DOWN

POWER

EXHAUST VALVE
OPEN

INTAKE VALVE
CLOSED

PISTON
MOVING UP

VALVE TAPPET
LIFTING VALVE

CAM LOBE LIFTING
VALVE TAPPET

EXHAUST

54.19

Figure 13-4.—Four-stroke cycle in a gasoline engine.

us see what happens during the actual running of the engine. To produce sustained power, an engine must accomplish a definite series of operations over and over again. All you have to do is follow one series of events—intake, compression, power, and exhaust—until they repeat themselves. This one series of events is called a cycle.

Most engines of today are called four-cycle engines. What is meant is four-stroke-cycle, but our habit of abbreviating has eliminated the middle word. Just the same, when you see four-cycle it means there are four strokes of the piston, two up and two down, to each cycle. Then it starts over again on another cycle of the same four strokes.

TWO-CYCLE ENGINE

In the two-cycle engine, the entire cycle of events (intake, compression, power, and exhaust) takes place in two piston strokes.

A two-cycle engine is shown in figure 13-5. Every other stroke in this engine is a power stroke. Each time the piston moves down it is on the power stroke. Intake, compression, power, and exhaust still take place, but they are completed in just two strokes. In figure 13-5 the intake and exhaust ports are cut into the cylinder wall instead of being placed at the top of the combustion chamber as in the four-cycle engine. As the piston moves down on its power stroke, it first uncovers the exhaust port to let

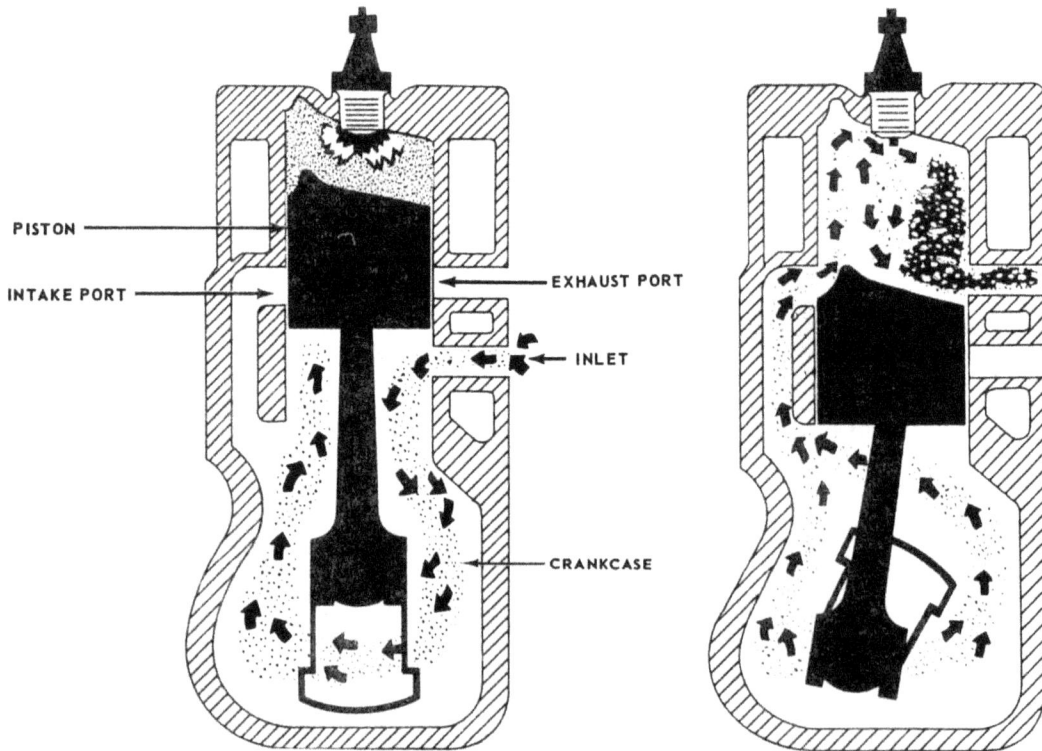

PISTON

INTAKE PORT

EXHAUST PORT

INLET

CRANKCASE

54.20

Figure 13-5.—Events in a two-cycle, internal combustion engine.

111

burned gases escape and then uncovers the intake port to allow a new fuel-air mixture to enter the combustion chamber. Then, on the upward stroke, the piston covers both ports and, at the same time, compresses the new mixture in preparation for ignition and another power stroke.

In the engine shown in figure 13-5 the piston is shaped so that the incoming fuel-air mixture is directed upward, thereby sweeping out ahead of it the burned exhaust gases. Also, there is an inlet into the crankcase through which the fuel-air mixture passes before it enters the cylinder. This inlet is opened as the piston moves upward, but it is sealed off as the piston moves downward on the power stroke. The downward moving piston slightly compresses the mixture in the crankcase, thus giving the mixture enough pressure to pass rapidly through the intake port as the piston clears this port. This improves the sweeping-out, or scavenging, effect of the mixture as it enters and clears the burned gases from the cylinder through the exhaust port.

FOUR-CYCLE vs TWO-CYCLE ENGINES

You have probably noted that the two-cycle engine produces a power stroke every crankshaft revolution; the four-cycle engine requires two crankshaft revolutions for each power stroke. It might appear then that the two-cycle could produce twice as much power as the four-cycle of the same size, operating at the same speed. However, this is not true. With the two-cycle engine some of the power is used to drive the blower that forces the air-fuel charge into the cylinder under pressure. Also, the burned gases are not completely cleared from the cylinder. Additionally, because of the much shorter period the intake port is open (as compared to the period the intake valve in a four-stroke-cycle is open), a relatively smaller amount of fuel-air mixture is admitted. Hence, with less fuel-air mixture, less power per power stroke is produced as compared to the power produced in a four-stroke cycle engine of like size operating at the same speed and with other conditions being the same. To increase the amount of fuel-air mixture, auxiliary devices are used with the two-stroke engine to ensure delivery of greater amounts of fuel-air mixture into the cylinder.

MULTIPLE-CYLINDER ENGINES

The discussion so far in this chapter has concerned a single-cylinder engine. A single cylinder provides only one power impulse every two crankshaft revolutions in a four-cycle engine and is delivering power only one-fourth of the time. To provide for a more continuous flow of power, modern engines use four, six, eight, or more cylinders. The same series of cycles take place in each cylinder.

In a four-stroke cycle six-cylinder engine, for example, the cranks on the crankshaft are set 120 degrees apart, the cranks for cylinders 1 and 6, 2 and 5, and 3 and 4 being in line with each other (fig. 13-6). The cylinders fire or deliver the power strokes in the following order: 1-5-3-6-2-4. Thus the power strokes follow each other so closely that there is a fairly continuous and even delivery of power to the crankshaft.

TIMING

In a gasoline engine, the valves must open and close at the proper times with regard to piston position and stroke. In addition, the ignition system must produce the sparks at the proper time so that the power strokes can start. Both valve and ignition system action must be properly timed if good engine performance is to be obtained.

Valve timing refers to the exact times in the engine cycle at which the valves trap the mixture and then allow the burned gases to escape. The valves must open and close so that they are constantly in step with the piston movement of the cylinder which they control. The position of the valves is determined by the camshaft; the position of the piston is determined by the crankshaft. Correct valve timing is obtained by providing the proper relationship between the camshaft and the crankshaft.

When the piston is at TDC the crankshaft can move 15° to 20° without causing the piston to move up and down any noticeable distance. This is one of the two rock positions (fig. 13-7). When the piston moves up on the exhaust stroke, considerable momentum is given to the exhaust gases as they pass out through the exhaust valve port, but if the exhaust valve closes at TDC, a small amount of the gases will be trapped and will dilute the incoming fuel-air mixture when the intake valves open. Since the piston has

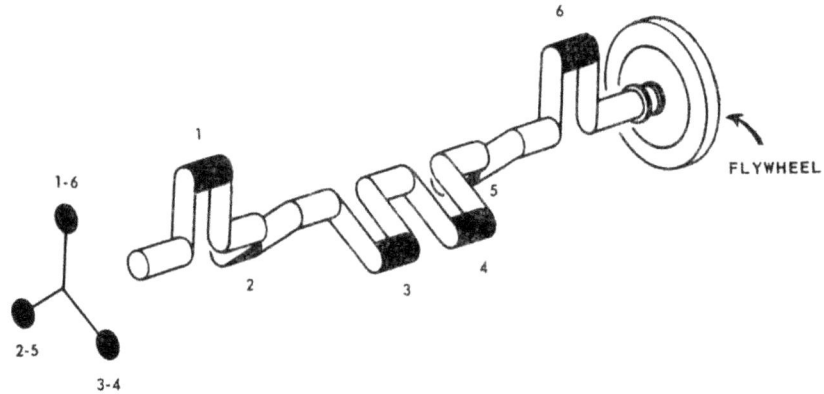

81.42

Figure 13-6.—Crankshaft for a six-cylinder engine.

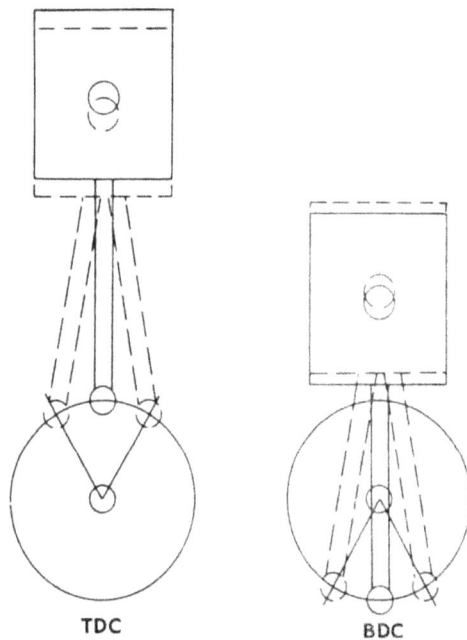

TDC   BDC

81.45

Figure 13-7.—Rock position.

little downward movement while in the rock position, the exhaust valve can remain open during this period and thereby permit a more complete scavenging of the exhaust gases.

Ignition timing refers to the timing of the spark at the spark plug gap with relation to the piston position during the compression and power strokes. The ignition system is timed so that the spark occurs before the piston reaches TDC on the compression stroke. This gives the mixture enough time to ignite and start burning. If this time were not provided, that is, if the spark occurred at or after TDC, then the pressure increase would not keep pace with the piston movement.

At higher speeds, there is still less time for the fuel-air mixture to ignite and burn. In order to compensate for this, and thereby avoid power loss, the ignition system includes an advance mechanism that functions on speed.

CLASSIFICATION OF ENGINES

Engines for automotive and construction equipment may be classified in a number of ways: type of fuel used; type of cooling employed; or valve and cylinder arrangement. They all operate on the internal combustion principle, and the application of basic principles of construction to particular needs or systems

113

of manufacture has caused certain designs to be recognized as conventional.

The most common method of classification is by the type of fuel used; that is, whether the engine burns gasoline or diesel fuel.

GASOLINE ENGINES VS DIESEL ENGINES

Mechanically and in overall appearance, gasoline and diesel engines resemble one another. However, in the diesel engine, many parts are somewhat heavier and stronger, so that they can withstand the higher temperatures and pressures the engine generates. The engines differ also in the fuel used, in the method of introducing it into the cylinders, and in how the air-fuel mixture is ignited. In the gasoline engine, air and fuel first are mixed together in the carburetor. After this mixture is compressed in the cylinders, it is ignited by an electrical spark from the spark plugs. The source of the energy producing the electrical spark may be a storage battery or a high-tension magneto.

The diesel engine has no carburetor. Air alone enters its cylinders, where it is compressed and reaches high temperature due to compression. The heat of compression ignites the fuel injected into the cylinder and causes the fuel-air mixture to burn. The diesel engine needs no spark plugs; the very contact of the diesel fuel with the hot air in the cylinders causes ignition. In the gasoline engine the heat from compression is not enough to ignite the air-fuel mixture, therefore spark plugs are necessary.

ARRANGEMENT OF CYLINDERS

Engines are classified also according to the arrangement of the cylinders: inline, with all cylinders cast in a straight line above the crankshaft, as in most trucks; and V-type with two banks of cylinders mounted in a "V" shape above the crankshaft, as in many passenger vehicles. Another not-so-common arrangement is the horizontally opposed engine whose cylinders are mounted in two side rows, each opposite a central crankshaft. Buses often are equipped with this type of engine.

The cylinders are numbered. The cylinder nearest the front of an in-line engine is No. 1. The others are numbered 2, 3, 4, etc., from front to rear. In V-type engines the numbering sequence varies with the manufacturer.

The firing order (which is different from the numbering order) of the cylinders is usually stamped on the cylinder block or on the manufacturer's nameplate.

VALVE ARRANGEMENT

The majority of internal combustion engines also are classified according to the position and arrangement of the intake and exhaust valves—that is, whether the valves are in the cylinder block or in the cylinder head. Various arrangements have been used, but the most common are L-head, I-head, and F-head (fig. 13-8). The letter designation is used because the shape of the combustion chamber resembles the form of the letter identifying it.

L-Head

In the L-head engines both valves are placed in the block on the same side of the cylinder. The valve-operating mechanism is located directly below the valves, and one camshaft actuates both the intake and exhaust valves.

I-Head

Engines using the I-head construction are commonly called valve-in-head or over-head valve engines, because the valves are mounted in a cylinder head above the cylinder. This arrangement requires a tappet, a push rod, and a rocker arm above the cylinder to reverse the direction of valve movement, but only one camshaft is required for both valves. Some overhead valve engines make use of an overhead camshaft. This arrangement eliminates the long linkage between the camshaft and valve.

F-Head

In the F-head engine, the intake valves normally are located in the head, while the exhaust valves are located in the engine block. This arrangement combines, in effect, the L-head and the I-head valve arrangements. The valves in the head are actuated from the camshaft through tappets, push rods, and rocker arms (I-head arrangement), while the valves in the block are actuated directly from the camshaft by tappets (L-head arrangement).

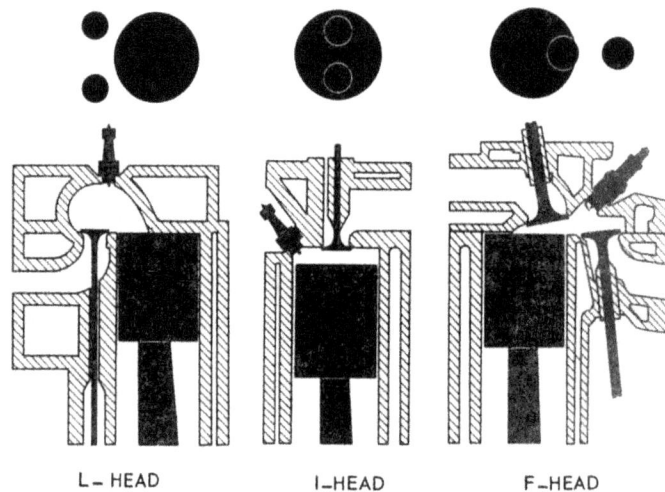

L – HEAD   I–HEAD   F –HEAD

81.46

Figure 13-8.—L-, I-, and F-valve arrangement.

ENGINE CONSTRUCTION

Basic engine construction varies little, regardless of size and design of the engine. The intended use of an engine must be considered before the design and size can be determined. The temperature at which an engine will operate has a great deal to do with determining what metals must be used in its construction.

To simplify the service parts problem in the field, and also to simplify servicing procedures, the present trend in engine construction and design is toward what is called engine families. There must, of necessity, be many different kinds of engines because there are many kinds of jobs to be done. However, the service and service parts problem can be simplified by designing engines so that they are closely related in cylinder size, valve arrangement, etc. As an example, the GM series 71 engines can be obtained in 2, 3, 4, and 6 cylinders; but they are so designed that the same pistons, connecting rods, bearings, valve operating mechanisms and valves can be used in all 4 engines.

Engine construction, in this chapter, will be broken down into two categories: stationary parts and moving parts.

STATIONARY PARTS

The stationary parts of an engine include the cylinder block, cylinders, cylinder head or heads, crankcase, and the exhaust and intake manifolds. These parts furnish the framework of the engine. All movable parts are attached to or fitted into this framework.

Engine Cylinder Block

The engine cylinder block is the basic frame of a liquid-cooled engine, whether it be in-line, horizontally-opposed, or V-type. The cylinder block and crankcase are often cast in one piece which is the heaviest single piece of metal in the engine. (See fig. 13-9). In small engines, where weight is an important consideration, the crankcase may be cast separately. In most large diesel engines, such as those used in power plants, the crankcase is cast separately and is attached to a heavy stationary engine base.

In practically all automotive and construction equipment, however, the cylinder block and crankcase are cast in one piece. In this course we are concerned primarily with liquid-cooled engines, of this type.

115

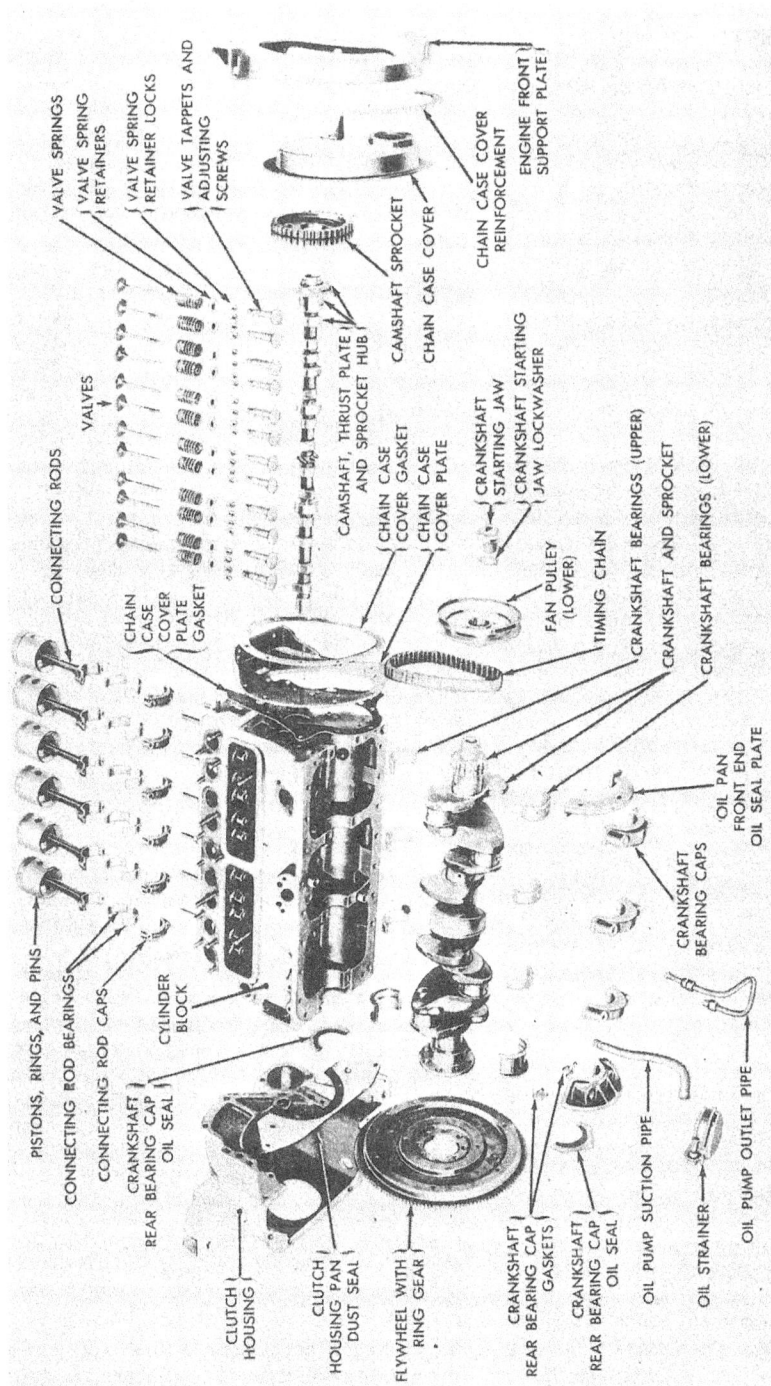

Figure 13-9.—Cylinder block and components.

81.47

The cylinders of a liquid-cooled engine are surrounded by jackets through which the cooling liquid circulates. These jackets are cast integrally with the cylinder block. Communicating passages permit the coolant to circulate around the cylinders and through the head.

The air-cooled engine cylinder differs from that of a liquid-cooled engine in that the cylinders are made individually, rather than cast in block. The cylinders of air-cooled engines have closely spaced fins surrounding the barrel; these fins provide a greatly increased surface area from which heat can be dissipated. This is in contrast to the liquid-cooled engine, which has a water jacket around its cylinders.

Cylinder Block Construction

The cylinder block is cast from gray iron or iron alloyed with other metals such as nickel, chromium, or molybdenum. Some light weight engine blocks are made from aluminum.

Cylinders are machined by grinding, and/or boring, to give them the desired true inner surface. During normal engine operation, cylinder walls will wear out-of-round, or they may become cracked and scored if not properly lubricated or cooled. Liners (sleeves) made of metal alloys resistant to wear, are used in many gasoline engines and practically all diesel engines to lessen wear. After they have been worn beyond the maximum oversize, the liners can by replaced individually permitting the use of standard pistons and rings. Thus you can avoid replacing the entire cylinder block.

The liners are inserted into a hole in the block with either a PRESS FIT or a SLIP FIT. Liners are further designated as WET TYPE or DRY TYPE. The wet type line comes in direct contact with the coolant and is sealed at the top by the use of a metallic sealing ring and at the bottom by a rubber sealing ring; the dry type liner does not contact the coolant.

Engine blocks for L-head engines contain the passageways for the valves and valve ports. The lower part of the block (crankcase) supports the crankshaft (with main bearings and bearing caps) and also provides a place for fastening the oil pan.

The camshaft is supported in the cylinder block by bushings that fit into machined holes in the block. On L-head in-line engines, the intake and exhaust manifolds are attached to the side of the cylinder block. On L-head V-8 engines, the intake manifold is located between the two banks

of cylinders. In this engine, there are two exhaust manifolds, one on the outside of each bank.

Cylinder Head

The cylinder head provides the combustion chambers for the engine cylinders. It is built to conform to the arrangement of the valves: L-head, I-head, or other.

In the water-cooled engine the cylinder head (fig. 13-10) is bolted to the top of the cylinder block to close the upper end of the cylinders. It contains passages, matching those of the cylinder block, which allow the cooling water to circulate in the head. The head also helps retain compression in the cylinders. In the gasoline engine there are tapped holes in the cylinder head which lead into the combustion chamber. The spark plugs are inserted into these tapped holes.

In the diesel engine the cylinder head may be cast in a single unit, or may be cast for a single cylinder or two or more cylinders. Separated head sections (usually covering 1, 2, or 3 cylinders in large engines) are easy to handle and can be readily removed.

The L-head type of cylinder head shown in figure 13-10 is a comparatively simple casting. It contains water jackets for cooling, and openings for spark plugs. Pockets into which the valves operate are also provided. Each pocket serves as a part of the combustion chamber. The fuel-air mixture is compressed in the pocket as the piston reaches the end of the compression stroke. Note that the pockets have a rather complex curved surface. This shape has been carefully designed so that the fuel-air mixture, in being compressed, will be subjected to violent turbulence. This turbulence assures uniform mixing of the fuel and air, thus improving the combustion process.

The I-head (overhead-valve) type of cylinder head contains not only water jackets for cooling spark-plug openings, and valve and combustion-chamber pockets, but it also contains and supports the valves and valve-operating mechanisms. In this type of cylinder head, the water jackets must be large enough to cool not only the top of the combustion chamber but also the valve seats, valves, and valve-operating mechanisms.

Crankcase

The crankcase is that part of the engine block below the cylinders. It supports and encloses the

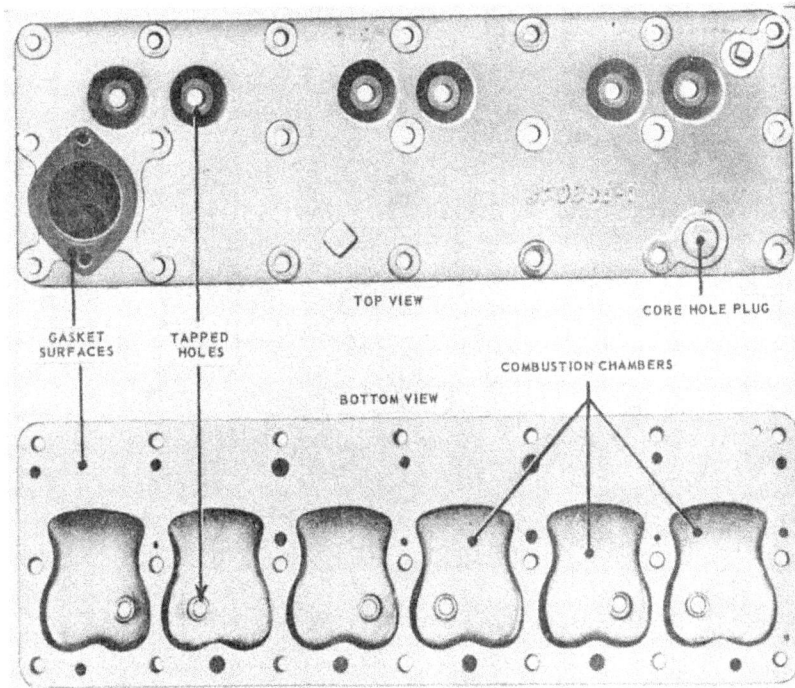

81.50

Figure 13-10.—Cylinder head for L-head engine.

crankshaft and provides a reservoir for the lubricating oil. Oftentimes there are places provided on the crankcase for the mounting of the oil pump, oil filter, starting motor, and the generator. The lower part of the crankcase is the OIL PAN, which is bolted at the bottom. The oil pan is made of pressed or cast steel and holds from 4 to 9 quarts of oil, depending on the engine design.

The crankcase also has mounting brackets which support the entire engine on the vehicle frame. These brackets are either an integral part of the crankcase or are bolted to it in such a way that they support the engine at 3 or 4 points. These points of contact usually are cushioned with rubber, which insulates the frame and body of the vehicle from engine vibration and therefore prevents damage to the engine supports and the transmission.

Exhaust Manifold

The exhaust manifold is essentially a tube that carries waste products of combustion from the cylinders. On L-head engines the exhaust manifold is bolted to the side of the engine block; on overhead-valve engines it is bolted to the side of the engine cylinder head. Exhaust manifolds may be single iron castings or may be cast in sections. They have a smooth interior surface with no abrupt changes in size. (See fig. 13-11.)

Intake Manifold

The intake manifold on a gasoline engine carries the fuel-air mixture from the carburetor and distributes it as evenly as possible to the cylinders. On a diesel engine the manifold carries only air to the cylinders. The intake

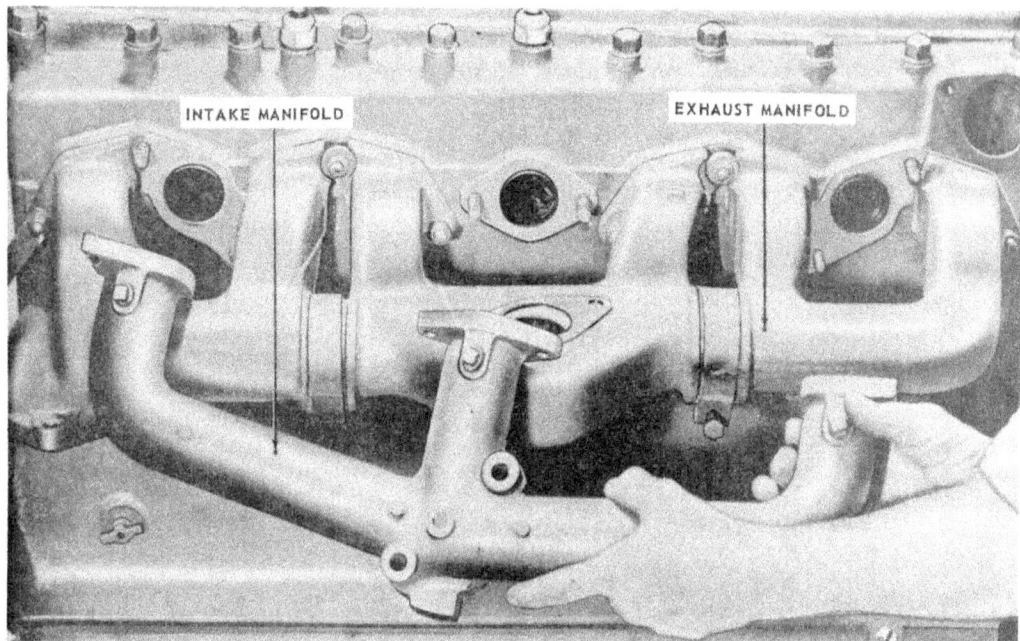

65.93

Figure 13-11.—Intake and exhaust manifolds.

manifold is attached to the block on L-head engines and to the side of the cylinder head on overhead-valve engines. (See fig. 13-11.)

In gasoline engines, smooth and efficient engine performance depends largely on whether or not the fuel-air mixtures that enter each cylinder are uniform in strength, quality, and degree of vaporization. The inside walls of the manifold must be smooth to offer little obstruction to the flow of the fuel-air mixture. The manifold is designed to prevent collecting of fuel at the bends in the manifold.

The intake manifold should be as short and straight as possible to reduce the chances of condensation between the carburetor and cylinders. To assist in vaporization of fuel, some intake manifolds are constructed so that part of their surfaces can be heated by hot exhaust gases.

Gaskets

The principal stationary parts of an engine have just been explained. The gaskets (fig. 13-12)

that serve as seals between these parts in assembly, require as much attention during assembly as any other part. It is impractical to machine all surfaces so that they fit together to form a perfect seal. The gaskets make a joint that is air, water, or oil tight; therefore, when properly installed, they prevent loss of compression, coolant, or lubricant.

MOVING PARTS OF AN ENGINE

The moving parts of an engine serve an important function in turning heat energy into mechanical energy. They further convert reciprocal motion into rotary motion. The principal moving parts are the piston assembly, connecting rods, crankshaft assembly (includes flywheel and vibration dampener), camshaft, valves, and gear train.

The burning of the fuel-air mixture within the cylinder exerts a pressure on the piston, thus pushing it down in the cylinder. The action of

119

the underside of the piston to reinforce the head; the ribs also assist in conducting heat from the head of the piston to the piston rings and out through the cylinder walls.

The structural components of the piston are the HEAD, SKIRT, RING GROOVES, and LANDS (fig. 13-14). However, all pistons do not look like the typical one here illustrated. Some have differently shaped heads. Diesel engine pistons usually have more ring grooves and rings than the pistons of gasoline engines. Some of these rings may be installed below as well as above the WRIST or PISTON PIN (fig. 13-15).

Fitting pistons properly is important. Because metal expands when heated, and because space must be provided for lubricants between the pistons and the cylinder walls, the pistons are fitted to the engine with a specified clearance. This clearance depends upon the size or diameter of the piston and the material from which it is made. Cast iron does not expand as fast or as much as aluminum. Aluminum pistons require more clearance to prevent binding or seizing

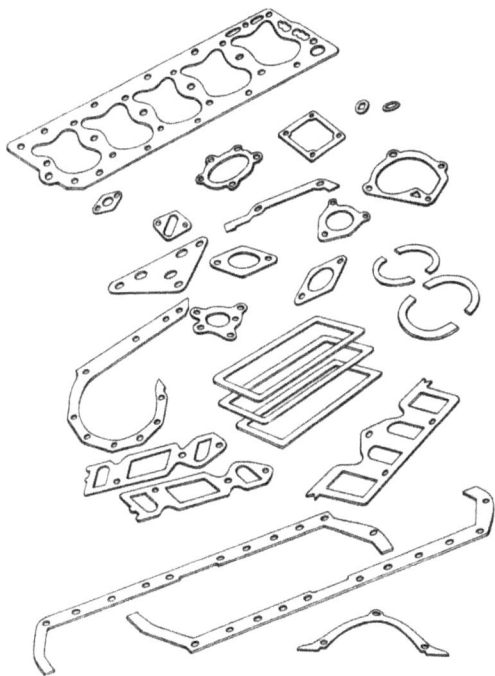

81.52

Figure 13-12.—Engine overhaul gasket kit.

the connecting rod and crankshaft converts this downward motion to a rotary motion.

Piston Assembly

Engine pistons serve several purposes: they transmit the force of combustion to the crankshaft through the connecting rod; they act as a guide for the upper end of the connecting rod; and they also serve as a carrier for the piston rings used to seal the compression in the cylinder. (See fig. 13-13.)

The piston must come to a complete stop at the end of each stroke before reversing its course in the cylinder. To withstand this rugged treatment and wear, it must be made of tough material, yet be light in weight. To overcome inertia and momentum at high speeds, it must be carefully balanced and weighed. All the pistons used in any one engine must be of similar weight to avoid excessive vibration. Ribs are used on

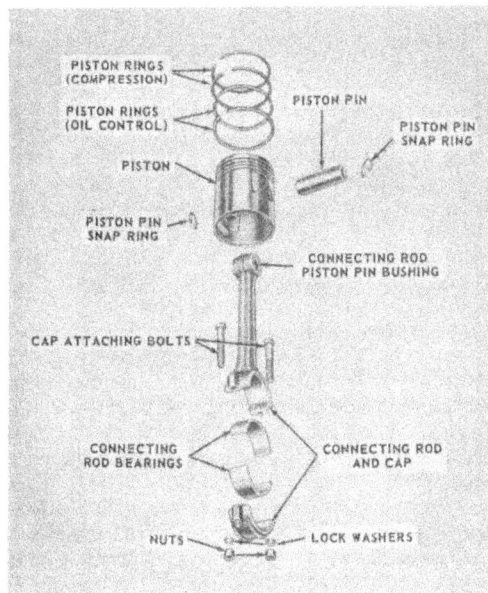

75.70

Figure 13-13.—Piston and connecting rod (exploded view).

120

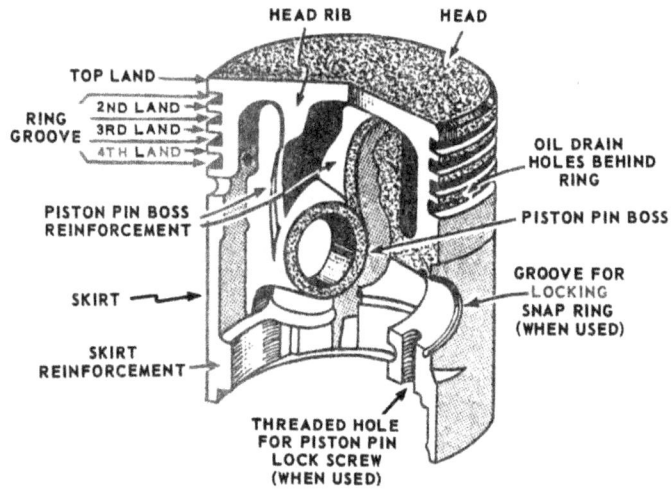

HEAD RIB   HEAD

TOP LAND

RING GROOVE
2ND LAND
3RD LAND
4TH LAND

OIL DRAIN HOLES BEHIND RING

PISTON PIN BOSS REINFORCEMENT

PISTON PIN BOSS

GROOVE FOR LOCKING SNAP RING (WHEN USED)

SKIRT

SKIRT REINFORCEMENT

THREADED HOLE FOR PISTON PIN LOCK SCREW (WHEN USED)

81.53

Figure 13-14.—The parts of a piston.

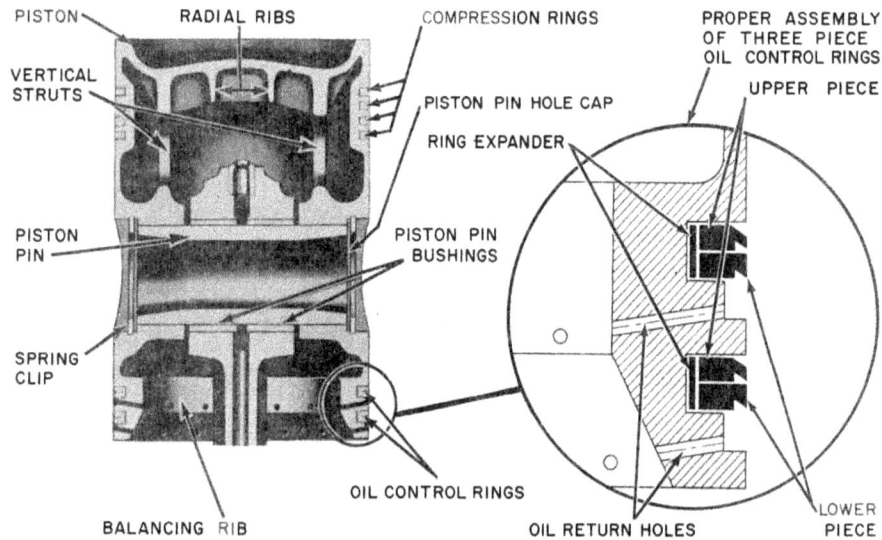

PISTON   RADIAL RIBS   COMPRESSION RINGS

PROPER ASSEMBLY OF THREE PIECE OIL CONTROL RINGS

UPPER PIECE

VERTICAL STRUTS

PISTON PIN HOLE CAP

RING EXPANDER

PISTON PIN

PISTON PIN BUSHINGS

SPRING CLIP

OIL CONTROL RINGS

BALANCING RIB

OIL RETURN HOLES

LOWER PIECE

81.54

Figure 13-15.—Piston assembly of General Motors series 71 engine.

121

when the engine gets hot. The skirt or bottom part of the piston runs much cooler than the top; therefore, it does not require as much clearance as the head.

The piston is kept in alignment by the skirt, which is usually CAM GROUND (elliptical in cross section) (fig. 13-16). This elliptical shape permits the piston to fit the cylinder, regardless of whether the piston is cold or at operating temperature. The narrowest diameter of the piston is at the piston pin bosses, where the metal is thickest. At the widest diameter of the piston, the piston skirt is thinnest. The piston is fitted to close limits at its widest diameter so that piston noise (slap) is prevented during engine warm-up. As the piston is expanded by the heat generated during operation, it becomes round because the expansion is proportional to the temperature of the metal. The walls of the skirt are cut away as much as possible to reduce weight and to prevent excessive expansion during

engine operation. Many aluminum pistons are made with SPLIT SKIRTS so that when the pistons expand the skirt diameter will not increase. The two types of piston skirts found in most engines are the FULL TRUNK and the SLIPPER. The full-trunk-type skirt, which is more widely used, has a full cylindrical shape with bearing surfaces parallel to those of the cylinder, giving more strength and better control of the oil film. The SLIPPER-TYPE (CUTAWAY) skirt has considerable relief on the sides of the skirt, leaving less area for possible contact with the cylinder walls and thereby reducing friction.

PISTON PINS.—The piston is attached to the connecting rod by means of the piston pin (wrist pin). The pin passes through the piston pin bosses and through the upper end of the connecting rod, which rides within the piston on the middle of the pin. Piston pins are made of alloy steel with a precision finish and are case

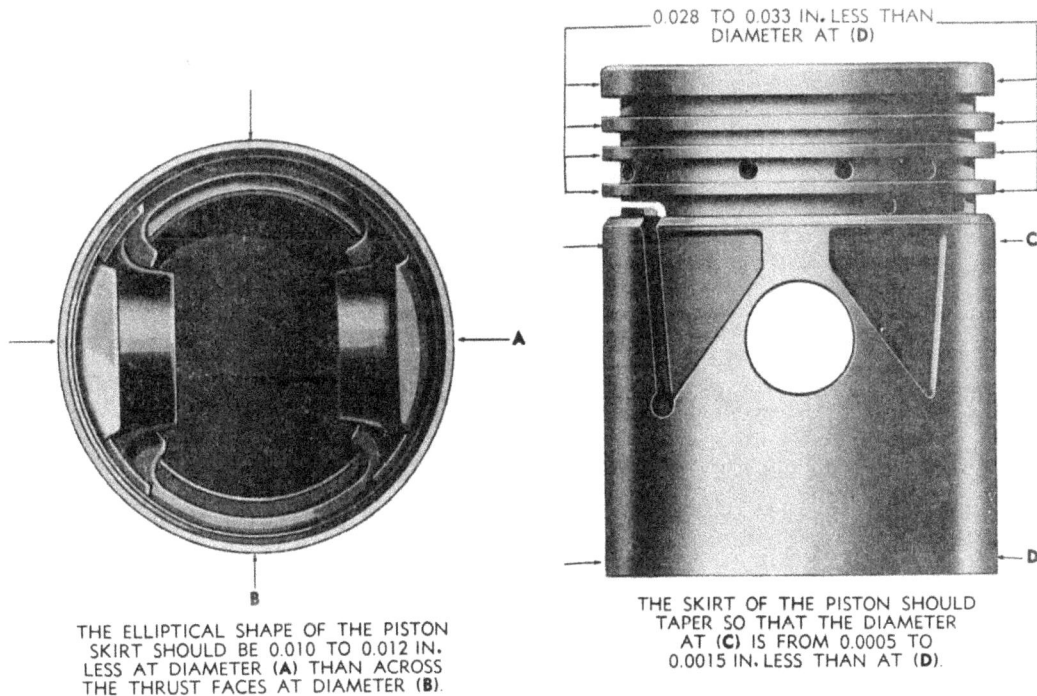

THE ELLIPTICAL SHAPE OF THE PISTON SKIRT SHOULD BE 0.010 TO 0.012 IN. LESS AT DIAMETER (A) THAN ACROSS THE THRUST FACES AT DIAMETER (B).

0.028 TO 0.033 IN. LESS THAN DIAMETER AT (D)

THE SKIRT OF THE PISTON SHOULD TAPER SO THAT THE DIAMETER AT (C) IS FROM 0.0005 TO 0.0015 IN. LESS THAN AT (D).

81.55

Figure 13-16.—Cam-ground piston.

hardened and sometimes chromium plated to increase their wearing qualities. Their tubular construction gives them a maximum of strength with a minimum of weight. They are lubricated by splash from the crankcase or by pressure through passages bored in the connecting rods.

There are three methods commonly used for fastening a piston pin to the piston and the connecting rod. (See fig. 13-17.) An anchored, or "fixed," pin is attached to the piston by a screw running through one of the bosses; the connecting rod oscillates on the pin. A "semifloating" pin is anchored to the connecting rod and turns in the piston pin bosses. A "full-floating" pin is free to rotate in the connecting rod and in the bosses, but is prevented from working out against the sides of the cylinder by plugs or snapring locks.

PISTON RINGS.—Piston rings are used on pistons to maintain gastight seals between the pistons and cylinders, to assist in cooling the piston, and to control cylinder-wall lubrication. About one-third of the heat absorbed by the piston passes through the rings to the cylinder wall. Piston rings are often quite complicated in design, are heat treated in various ways and are plated with other metals. There are two distinct classifications of piston rings: compression rings and oil control rings. (See fig. 13-18.)

The principal function of a compression ring is to prevent gases from leaking by the piston during the compression and power strokes. All piston rings are split to permit assembly to the piston and to allow for expansion. When the ring is in place, the ends of the split joint do not form a perfect seal; therefore, it is necessary to use more than one ring and to stagger the joints around the piston. If cylinders are worn, expanders (fig. 13-15 and 13-18) are sometimes used to ensure a perfect seal.

The bottom ring, usually located just above the piston pin, is an oil regulating ring. This ring scrapes the excess oil from the cylinder walls and returns some of it, through slots, to the piston ring grooves. The ring groove under an oil ring is provided with openings through which the oil flows back into the crankcase. In some engines, additional oil rings are used in the piston skirt below the piston pin.

Connecting Rods

Connecting rods must be light and yet strong enough to transmit the thrust of the pistons to the crankshaft. Connecting rods are drop forged from a steel alloy capable of withstanding heavy loads without bending or twisting. Holes at the upper and lower ends are machined to permit accurate fitting of bearings. These holes must be parallel.

The upper end of the connecting rod is connected to the piston by the piston pin. If the piston pin is locked in the piston pin bosses, or if it floats in both piston and connecting rod, the upper hole of the connecting rod will have a solid bearing (bushing) of bronze or similar material. As the lower end of the connecting rod revolves with the crankshaft, the upper end is forced to

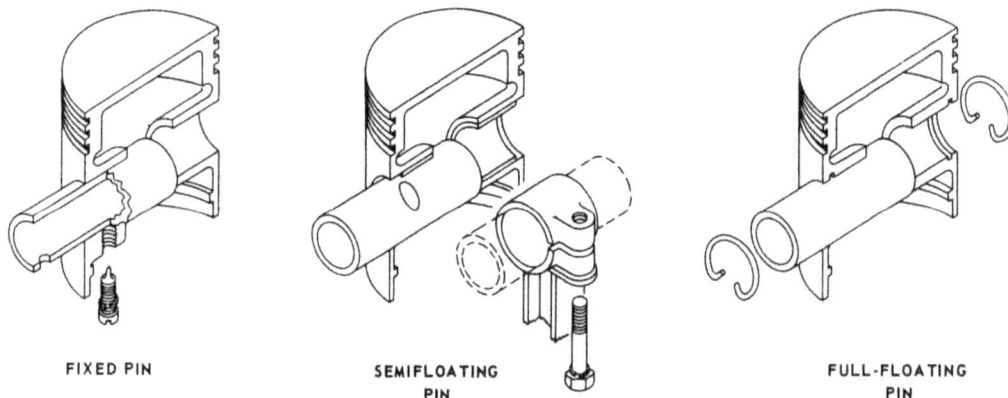

FIXED PIN   SEMIFLOATING PIN   FULL-FLOATING PIN

81.56

Figure 13-17.—Piston pin types.

The PRECISION type bearing is accurately finished to fit the crankpin and does not require further fitting during installation. It is positioned by projections on the shell which match reliefs in the rod and cap. The projections prevent the bearings from moving sideways and from rotary motion in the rod and cap.

The SEMIPRECISION type bearing is usually fastened to or die-cast with the rod and cap. Prior to installation, it is machined and fitted to the proper inside diameter with the cap and rod bolted together.

Crankshaft

As the pistons collectively might be regarded as the heart of the engine, so the CRANKSHAFT may be considered its backbone (fig. 13-19). It ties together the reactions of the pistons and the connecting rods, transforming their reciprocating motion into a rotary motion. And it transmits engine power through the flywheel, clutch, transmission, and differential to drive your vehicle.

The crankshaft is forged or cast from an alloy of steel and nickel, is machined smooth to provide bearing surfaces for the connecting rods and the main bearings, and is CASE-HARDENED, or coated in a furnace with copper alloyed with carbon. These bearing surfaces are called JOURNALS. The crankshaft counterweights impede the centrifugal force of the connecting rod assembly attached to the THROWS or points of bearing support. These

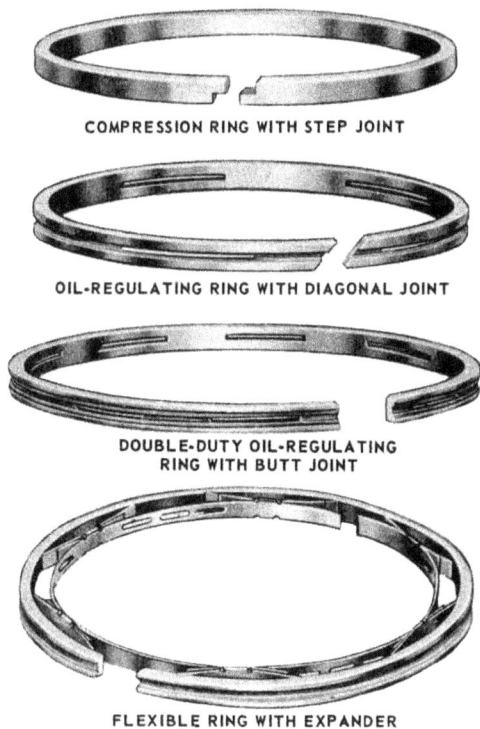

COMPRESSION RING WITH STEP JOINT

OIL-REGULATING RING WITH DIAGONAL JOINT

DOUBLE-DUTY OIL-REGULATING RING WITH BUTT JOINT

FLEXIBLE RING WITH EXPANDER

75.51

Figure 13-18.—Piston rings.

turn back and forth on the piston pin. Although this movement is slight, the bushing is necessary because the temperatures and the pressures are high. If the piston pin is semifloating, a bushing is not needed.

The lower hole in the connecting rod is split to permit it to be clamped around the crankshaft. The bottom part, or cap, is made of the same material as the rod and is attached by two or more bolts. The surface that bears on the crankshaft is generally a bearing material in the form of a separate split shell, although, in a few cases, it may be spun or die-cast in the inside of the rod and cap during manufacture. The two parts of the separate bearing are positioned in the rod and cap by dowel pins, projections, or short brass screws. Split bearings may be of the precision or semiprecision type.

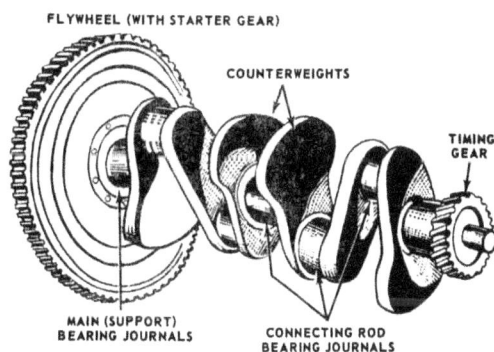

Figure 13-19.—Crankshaft of a 4-cylinder engine.

75.81

throws must be placed so that they counter-balance each other.

Crank throw arrangements for 4-, 6-, and 8-cylinder engines are shown in figure 13-20. Four-cylinder engine crankshafts have either 3 or 5 main support bearings and 4 throws in one plane. In figure 13-20 you see that the throws for No. 1 and No. 4 cylinders (4-cylinder engine) are 180° from those for No. 2 and No. 3 cylinders. On 6-cylinder engine crankshafts each of the 3 pairs of throws is arranged 120° from the other 2. Such crankshafts may be supported by as many as 7 main bearings, that is one at each end of the shaft and one between each pair of crankshaft throws. The crank-shafts of 8-cylinder V-type engines are similar to those for the 4-cylinder in-line type or may have each of the 4 throws fixed at 90° from each other (as in fig. 13-20) for better balance and smoother operation.

V-type engines usually have two connecting rods fastened side by side on one crankshaft throw. With this arrangement, one bank of the engine cylinders is set slightly ahead of the other to allow the two rods to clear each other.

Vibration Damper

The power impulses of an engine tend to set up torsional vibration in the crankshaft. If this torsional vibration were not controlled, the crankshaft might actually break at certain speeds; a vibration damper mounted on the front of the crankshaft is used to control this vibration (fig. 13-21).

Most types of vibration dampers resemble a miniature clutch. A friction facing is mounted between the hub face and a small damper fly-wheel. The damper flywheel is mounted on the hub face with bolts that go through rubber cones in the flywheel. These cones permit limited circumferential movement between the crank-shaft and damper flywheel. This minimizes the

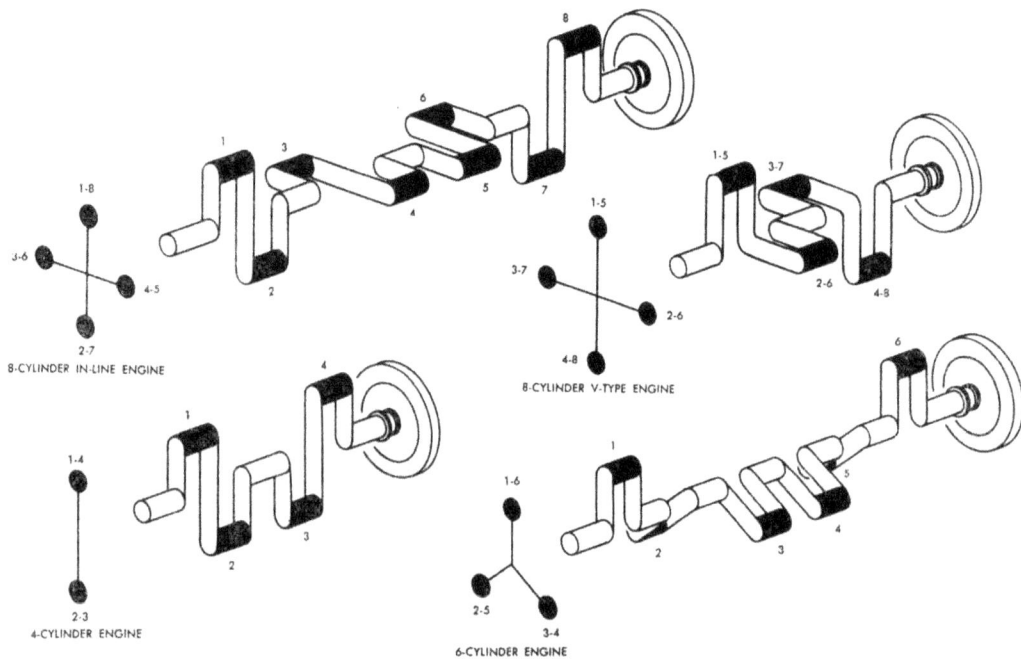

75.81

Figure 13-20.—Crankshaft and throw arrangements commonly used.

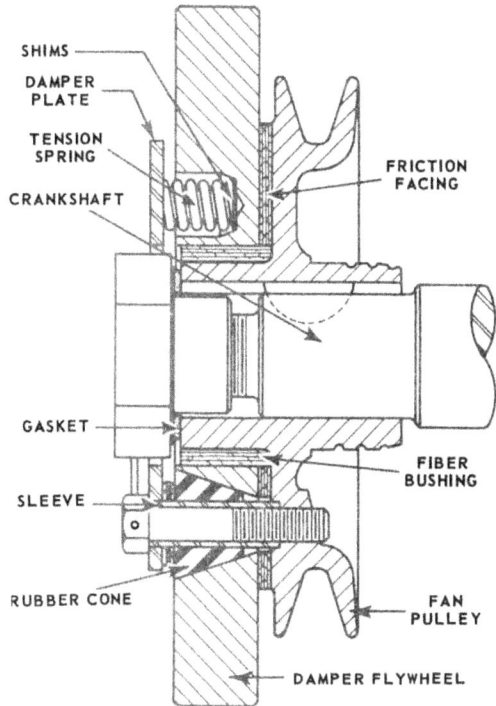

81.62

Figure 13-21.—Sectional view of a typical vibration damper.

effects of the torsional vibration in the crank-shaft. Several other types of vibration dampers are used. However, they all operate in essentially the same way.

Engine Flywheel

The flywheel is mounted at the rear of the crankshaft near the rear main bearing. This is usually the longest and heaviest main bearing in the engine, as it must support the weight of the flywheel.

The flywheel (fig. 13-22) stores up energy of rotation during power impulses of the engine. It releases this energy between power impulses, thus assuring less fluctuation in engine speed and smoother engine operation. The size of the flywheel will vary with the number of cylinders and the general construction of the engine. With a large number of cylinders and the consequent

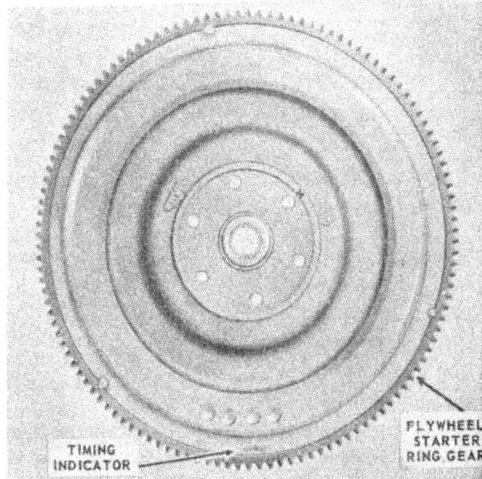

81.63

Figure 13-22.—Flywheel.

overlapping of power impulses, there is less need for a flywheel; consequently, the flywheel can be relatively small. The flywheel rim carries a ring gear, either integral with the fly-wheel or shrunk on, that meshes with the starter driving gear for cranking the engine. The rear face of the flywheel is usually machined and ground, and acts as one of the pressure surfaces for the clutch, becoming a part of the clutch assembly.

Valves And Valve Mechanisms

There are two valves for each cylinder in most engines, one intake and one exhaust valve. Since each of these valves operates at different times, it is necessary that separate operating mechanisms be provided for each valve. Valves are normally held closed by heavy springs and by compression in the combustion chamber. The purpose of the valve-actuating mechanism is to overcome the spring pressure and open the valves at the proper time. The valve-actuating mechanism includes the engine camshaft, cam-shaft followers (tappets), pushrods, and rocker arms.

CAMSHAFT.—The camshaft (fig. 13-23) is inclosed in the engine block. It has eccentric lobes (cams) ground on it for each valve in the

126

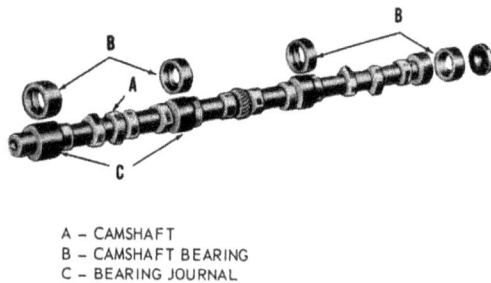

A – CAMSHAFT
B – CAMSHAFT BEARING
C – BEARING JOURNAL

81.61

Figure 13-23.—Camshaft and bushings.

engine. As the camshaft rotates, the cam lobe moves up under the valve tappet, exerting an upward thrust through the tappet against the valve stem or a pushrod. This thrust overcomes the valve spring pressure as well as the gas pressure in the cylinder, causing the valve to open. When the lobe moves from under the tappet, the valve spring pressure reseats the valve.

On L-, F-, or I-head engines, the camshaft is usually located to one side and above the crankshaft, while in V-type engines it is usually located directly above the crankshaft. On the overhead camshaft engine, such as the Murphy diesel, the camshaft is located above the cylinder head.

The camshaft of a 4-stroke cycle engine turns at one-half engine speed. It is driven off the crankshaft through timing gears or a timing chain. In the 2-stroke cycle engine the camshaft must turn at the same speed as the crankshaft in order that each valve may open and close once in each revolution of the engine.

In most cases the camshaft will do more than operate the valve mechanism. It may have extra cams or gears that operate fuel pumps, fuel injectors, the ignition distributor, or the lubrication pump.

Camshafts are supported in the engine block by journals in bearings. Camshaft bearing journals are the largest machined surfaces on the shaft. The bearings are usually made of bronze and are bushings rather than split bearings. The bushings are lubricated by oil circulating through drilled passages from the crankcase. The stresses on the camshaft are

small, therefore the bushings are not adjustable and require little attention. The camshaft bushings are generally replaced only when the engine requires a complete overhaul.

FOLLOWERS.—Camshaft followers (figs. 13-24 and 13-25) are the parts of the valve-actuating mechanism that contact the camshaft. You will probably hear them called valve tappets or valve lifters. In the L-head engine the followers directly contact the end of the valve stem and have an adjusting device in them. In the overhead valve engine the followers contact the pushrod that operates the rocker arm. The end of the rocker arm opposite the pushrod contacts the valve stem. The valve adjusting device, in this case, is in the rocker arm.

Many engines have self-adjusting valve lifters of the hydraulic type that operate at zero clearance at all times. The operation of one type of hydraulic valve tappet mechanism is shown in figure 13-26. Oil under pressure is forced into the tappet when the valve is closed, and this pressure extends the plunger in the tappet so that all valve clearance, or lash, is eliminated. When the cam lobe moves around under the tappet and starts to raise it, there will not be any tappet noise. As the lobe starts to

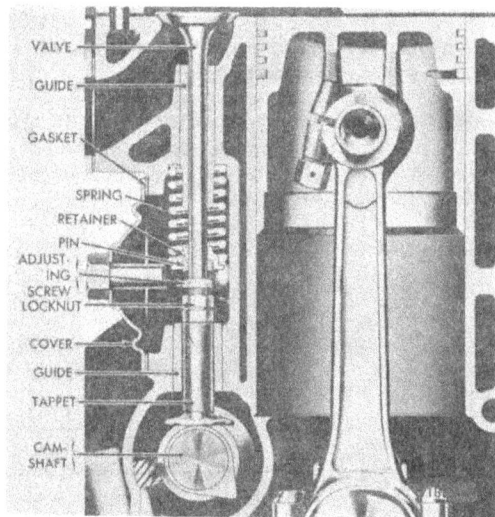

81.64

Figure 13-24.—L-head valve operating mechanism.

127

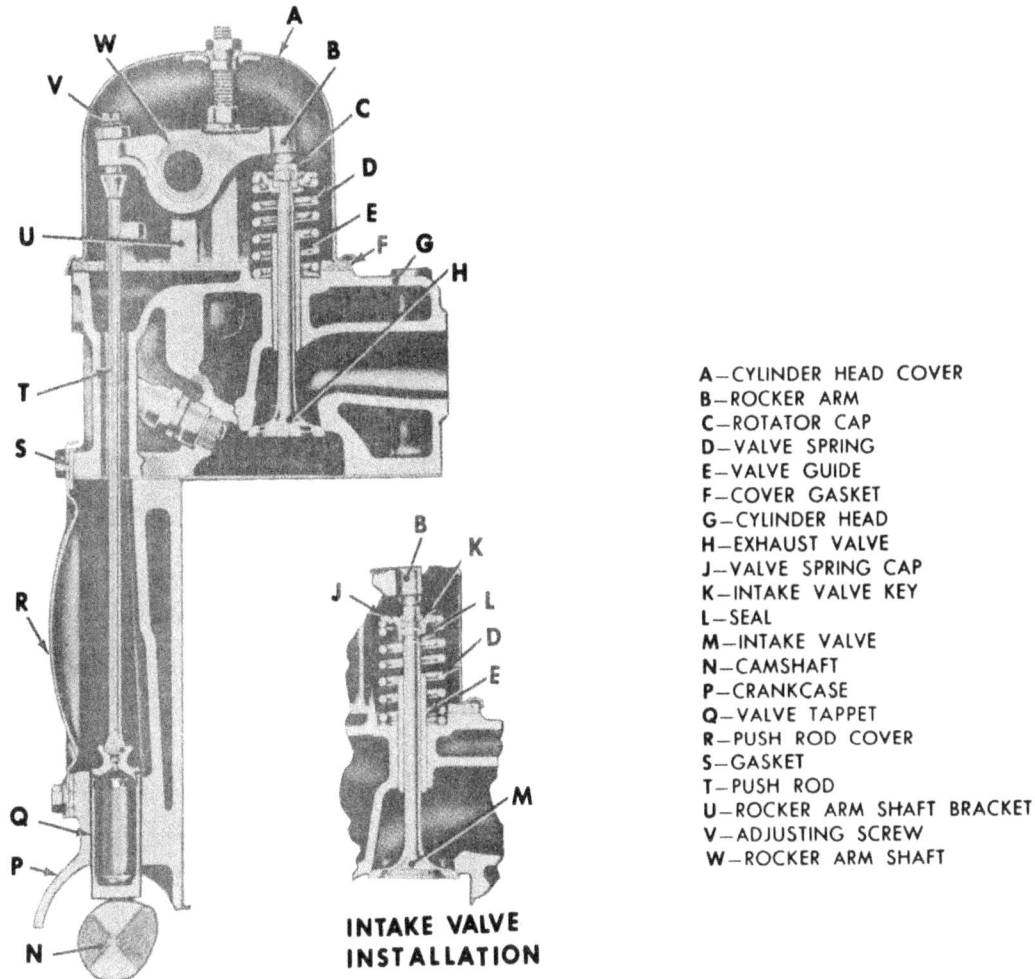

A—CYLINDER HEAD COVER
B—ROCKER ARM
C—ROTATOR CAP
D—VALVE SPRING
E—VALVE GUIDE
F—COVER GASKET
G—CYLINDER HEAD
H—EXHAUST VALVE
J—VALVE SPRING CAP
K—INTAKE VALVE KEY
L—SEAL
M—INTAKE VALVE
N—CAMSHAFT
P—CRANKCASE
Q—VALVE TAPPET
R—PUSH ROD COVER
S—GASKET
T—PUSH ROD
U—ROCKER ARM SHAFT BRACKET
V—ADJUSTING SCREW
W—ROCKER ARM SHAFT

INTAKE VALVE
INSTALLATION

81.65

Figure 13-25.—Valve operating mechanism for an overhead valve engine.

raise the tappet, the oil is forced upward in the lower chamber of the tappet. This action closes the ball check valve so oil cannot escape. Then the tappet acts as though it were a simple, 1-piece tappet and the valve is opened. When the lobe moves out from under the tappet and the valve closes, the pressure in the lower chamber of the tappet is relieved. Any slight loss of oil from the lower chamber is then replaced by the oil pressure from the engine lubricating system. This causes the plunger to move up snugly against the push rod so that any clearance is eliminated.

Timing Gears (Gear Trains)

Timing gears keep the crankshaft and camshaft turning in proper relation to one another so that the valves open and close at the proper time. In some engines, sprockets and chains are used.

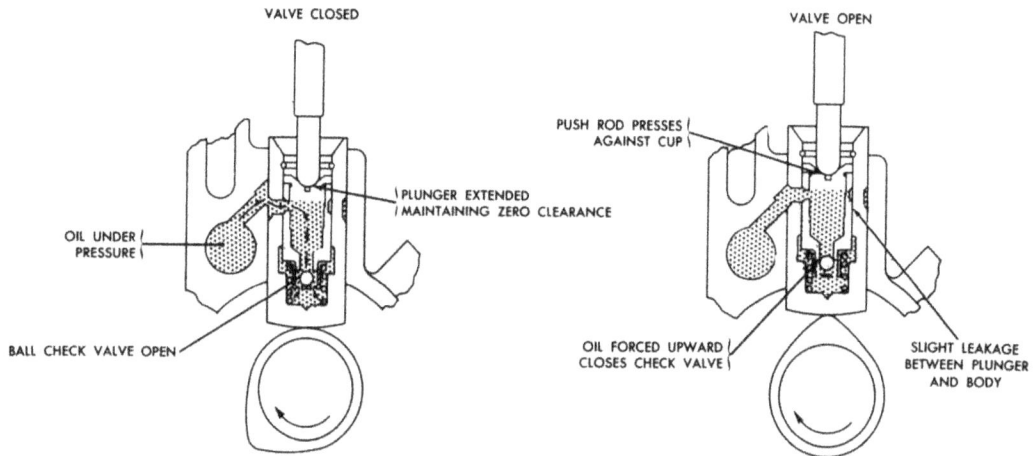

81.66

Figure 13-26.—Operation of a hydraulic valve lifter.

The gears or sprockets, as the case may be, of the camshaft and crankshaft are keyed in position so that they cannot slip. Since they are keyed to their respective shafts, they can be replaced if they become worn or noisy.

With directly driven timing gears (fig. 13-27), one gear usually has a mark on two adjacent teeth and the other a mark on only one tooth. To time the valves properly, it is necessary only to mesh the gears so that the two marked teeth of one gear straddle the single marked tooth of the other.

AUXILIARY ASSEMBLIES

We have discussed the main parts of the engine proper; but there are other parts, both moving and stationary, that are essential to engine operation. They are not built into the engine itself, but, in most cases, are attached to the engine block or cylinder head.

The fuel system includes a fuel pump and carburetor mounted on the engine. In diesel engines the fuel injection mechanism replaces the carburetor. An electrical system is provided to supply power for starting the engine and also for ignition during operation. An efficient cooling system is necessary for operating an internal combustion engine. In water-cooled engines a water pump and fan are used, while in air-cooled engines a blower is generally used to force cool air around the engine cylinders.

In addition, an exhaust system is provided to carry away the burned gases exhausted from the engine cylinders. These systems will not be discussed in this course, however. For further information on them refer to NavPers 10644D, Construction Mechanic 3 & 2.

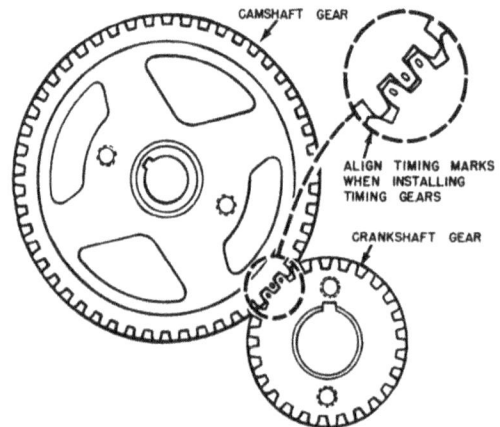

81.69

Figure 13-27.—Timing gears and their markings.

129

CHAPTER 14

POWER TRAINS

In chapter 13 we saw how a combination of simple machines and basic mechanisms were utilized in constructing the internal combustion engine. In this chapter we will go on from there to learn how the power developed by the engine is transmitted to perform the work required of it. To illustrate this, we will use the power train system as used by the automobile, and most trucks, as a familiar example. In this application, once again you are to look for the simple machines that make up each of the machines or mechanisms which are interconnected to make up the power train.

In a vehicle, the mechanism that transmits the power of the engine to the wheels and/or tracks and accessory equipment is called the power train. In a simple situation, a set of gears or a chain and sprocket could perform this task, but automotive and construction vehicles are not usually designed for such simple operating conditions. They are designed to have great pulling power, to move at high speeds, to travel in reverse as well as forward, and to operate on rough terrain as well as smooth roads. To meet these widely varying demands, a number of units have been added to the vehicles.

Automobiles and light trucks driven by the two rear wheels have a power train consisting of clutch, transmission, propeller shaft, differential, and driving axles (fig. 14-1).

In 4- and 6-wheel drive trucks, you will find transfer cases with additional drive shafts and live axles. Tractors, shovels, cranes, and other heavy-duty vehicles that move on tracks also have similar power trains. In addition to assemblies that drive sprockets to move the tracks, these vehicles also have auxiliary transmissions or power takeoff units which may be used to operate accessory attachments. The propeller shafts and clutch assemblies of these power trains are very much like those used to drive the wheels.

THE CLUTCH

The clutch is placed in the power train of motorized equipment for two purposes:

First, it provides a means of disconnecting the power of the engine from the driving wheels and accessory equipment. When the clutch is disengaged, the engine can run without driving the vehicle or operating the accessories.

Second, when the vehicle is started, the clutch allows the engine to take up the load of driving the vehicle or accessories gradually and without shock.

Clutches are located in the power train between the source of power and the operating unit. Usually, they are placed between the engine and the transmission assembly, as shown in figure 14-1.

Clutches generally transmit power from the clutch driving member to the driven member by friction. In the plate clutch, figure 14-2 the driving member or plate, which is secured to the engine flywheel, is gradually brought in contact with the driven member (disc). The contact is made and held by strong spring pressure controlled by the driver with the clutch pedal. With only a light spring pressure, there is little friction between the two members and the clutch is permitted to slip. As the spring pressure increases, friction also increases, and less slippage occurs. When the driver removes his foot from the clutch pedal and full spring pressure is applied, the speed of the driving plate and driven disc is the same, and all slipping stops. There is then a direct connection between the driving and driven shafts.

In most clutches, there is a direct mechanical linkage between the clutch pedal and the clutch release yoke lever. On many late model vehicles, and on some of the larger units which require great pressure to release the spring, a hydraulic clutch release system is used. A master cylinder (fig. 14-3), similar to the brake

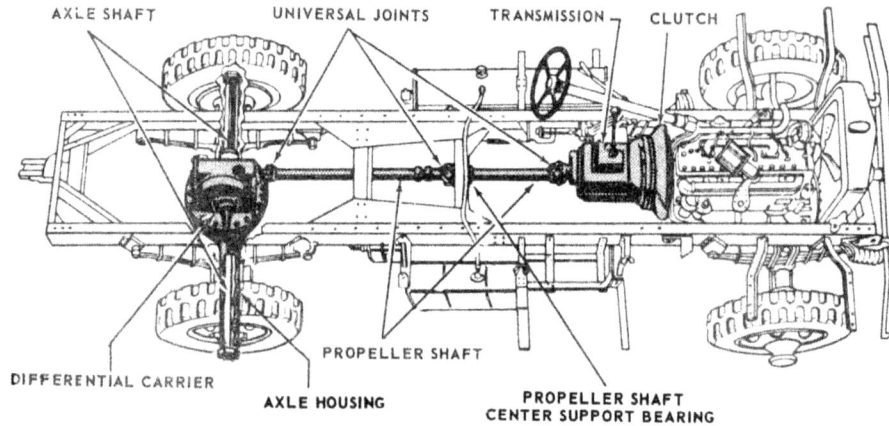

81.174

Figure 14-1.—Type of power transmission.

master cylinder, is attached to the clutch pedal. A cylinder, similar to a single-acting brake wheel cylinder, is connected to the master cylinder by flexible pressure hose or metal tubing (fig. 14-3). The slave cylinder is connected to the clutch release yoke lever. Movement of the clutch pedal actuates the clutch master cylinder. This movement is transferred by hydraulic pressure to the slave cylinder, which in turn actuates the clutch release yoke lever.

TYPES OF CLUTCHES

There are various types of clutches. The type most used in passenger cars and light

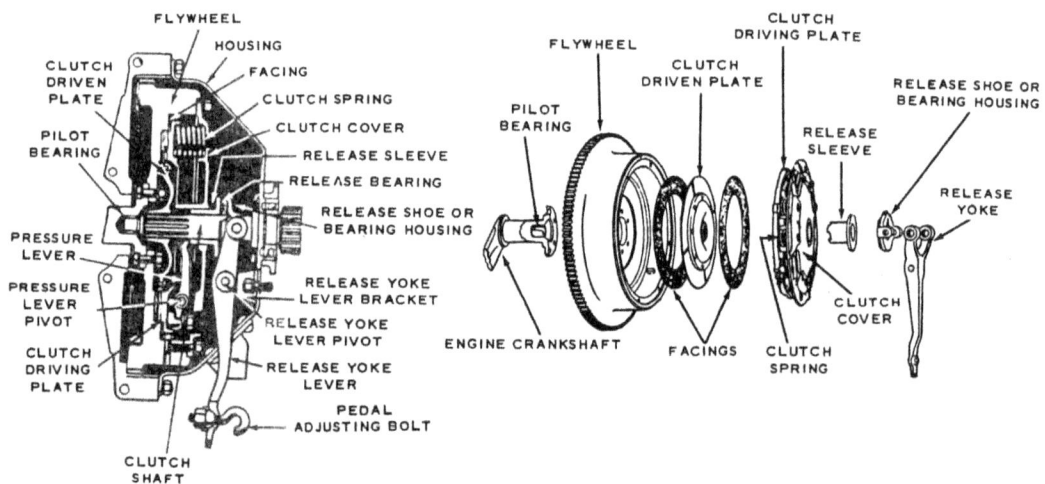

81.175

Figure 14-2.—Exploded and cross-section view of a plate clutch.

131

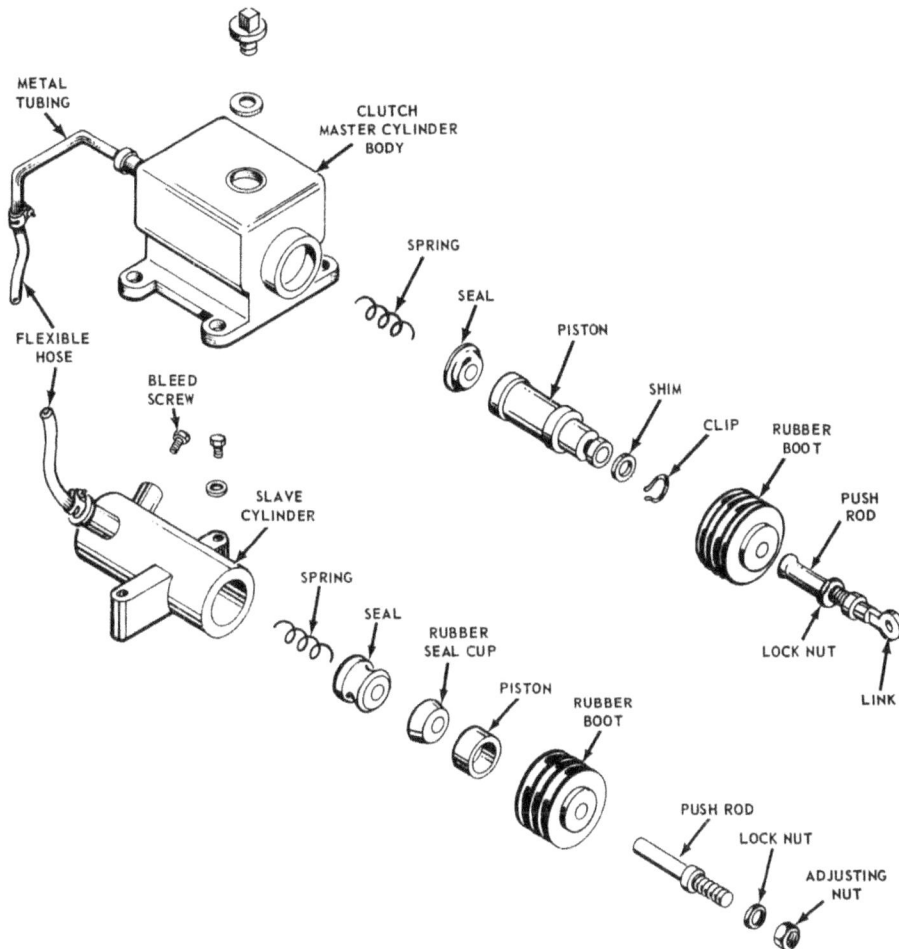

81.177

Figure 14-3.—Master cylinder, slave cylinder and connections for standard hydraulic clutch.

trucks is the previously-mentioned plate clutch. The plate clutch is a simple clutch with three plates, one of which is clamped between the other two. Exploded and cross-sectional views of a plate clutch are shown in figure 14-2.

Single Disk Clutch

The driving members of the single disk clutch consist of the flywheel and the driving (pressure) plate. The driven member consists of a single disk, splined to the clutch shaft and faced on both sides with friction material. When the clutch is fully engaged, the driven disc is firmly clamped between the flywheel and the driving plate by pressure of the clutch springs, forming a direct, nonslipping connection between the driving and driven members of the clutch. In this position, the driven disc rotates the clutch shaft to which it is splined. The clutch shaft is

132

connected to the driving wheels through the transmission, propeller shaft, final drive, differential, and live axles.

The double disk clutch (fig. 14-4) is substantially the same as the single plate disk clutch except that another driven disk and intermediate driving plate is added.

Multiple Disk Clutch

A multiple disk clutch is one having more than three plates or disks. Some have as many as 11 driving plates and 10 driven disks. Because the multiple disk type has a greater frictional area than a plate clutch, it is best suited as a steering clutch on crawler type tractors. The multiple disk clutch is sometimes used on heavy trucks. In operation, it is very much like the plate clutch and has the same release mechanism. The facings, however, are usually attached to the driving plates rather than to the driven disks. This reduces the weight of the driven disks and keeps them from spinning after the clutch is released.

You may run into other types of friction clutches such as the lubricated plate clutch and the cone clutch. These types are seldom used on automotive equipment. However, fluid drive is largely replacing the friction clutches in automobiles and light trucks, and even in some tractors.

For information on fluid drives (automatic transmissions), refer to Construction Mechanic 3 & 2, NavPers 10644-D, chapter 11.

TRANSMISSION

The transmission is part of the power train. It consists of a metal case filled with gears (fig. 14-5), and is usually located in the rear of the engine between the clutch housing and the propeller shaft, as shown in figure 14-1. The transmission transfers engine power from the clutch shaft to the propeller shaft, and allows the driver or operator to control the power and speed of the vehicle. The transmission shown in figure 14-5 and 14-6 is a sliding gear transmission. Many late model trucks

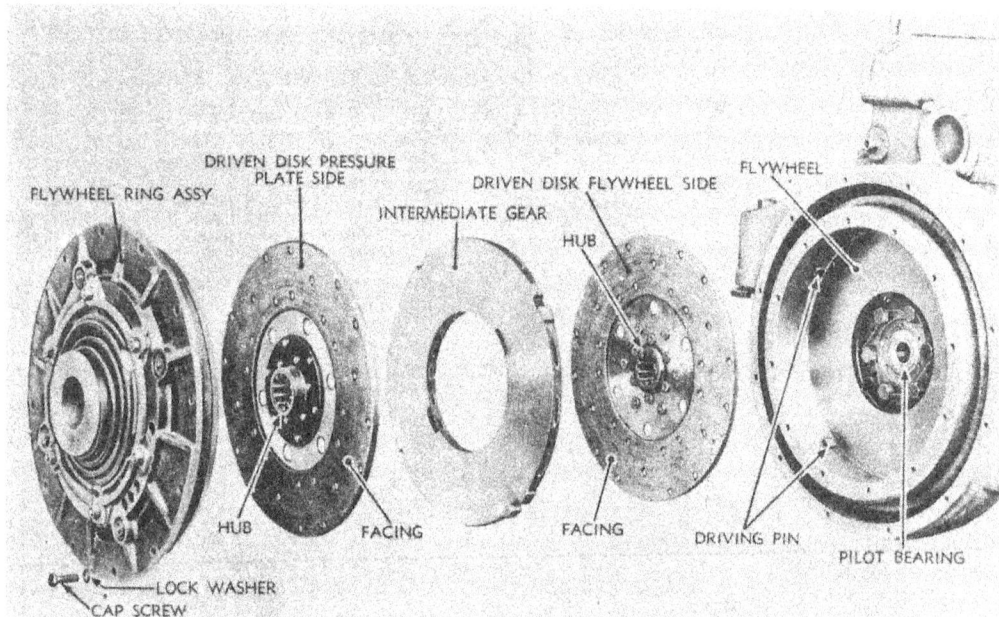

81.179

Figure 14-4.—Double disk clutch—exploded view.

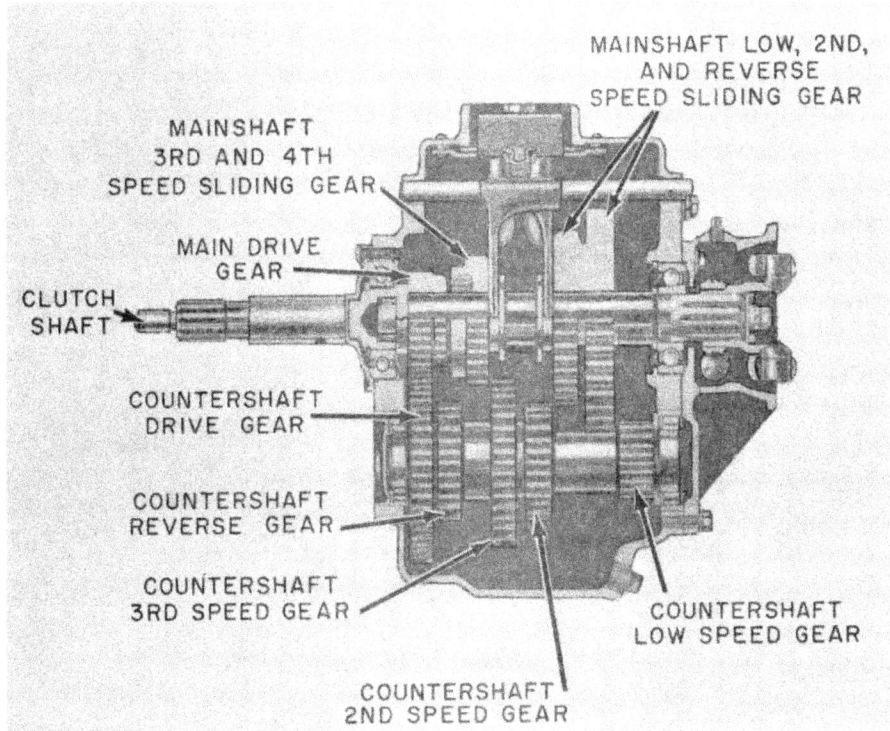

81.181

Figure 14-5.—Four-speed truck transmission.

have either constant mesh or synchromesh transmissions (explained later). However, the principles of operation and gear ratios are the same.

A review of chapter 6 of this book will help you to understand the transmissions and power transfer mechanisms described in this chapter.

FOUR-SPEED TRUCK TRANSMISSION

The gear shift lever positions shown in the small inset in figure 14-6 are typical of most four-speed truck transmissions. The gear shifting lever, shown at A, B, C, D, and E in the illustration, moves the position of the two shifting forks which slide on separate shafts secured in the transmission case cover. Follow the separate diagrams to learn what takes place in shifting from one speed to another. For example, as you

move the top of the gear shift lever toward the forward left position, the lower arm of the lever moves in the opposite direction to shift the gears. The fulcrum of this lever is in the transmission cover.

In shifting transmission gears it is necessary to use the clutch to disengage the engine. Improper use of the clutch will cause the gears to clash, and may damage them by breaking the gear teeth. A broken tooth or piece of metal can wedge itself between two moving gears and ruin the entire transmission assembly.

When you shift from neutral to first or low speed (A of fig. 14-6), the smallest countershaft gear engages with the largest sliding gear. Low gear moves the truck at its lowest speed and maximum power. The arrow indicates the flow of power from the clutch shaft to the propeller shaft.

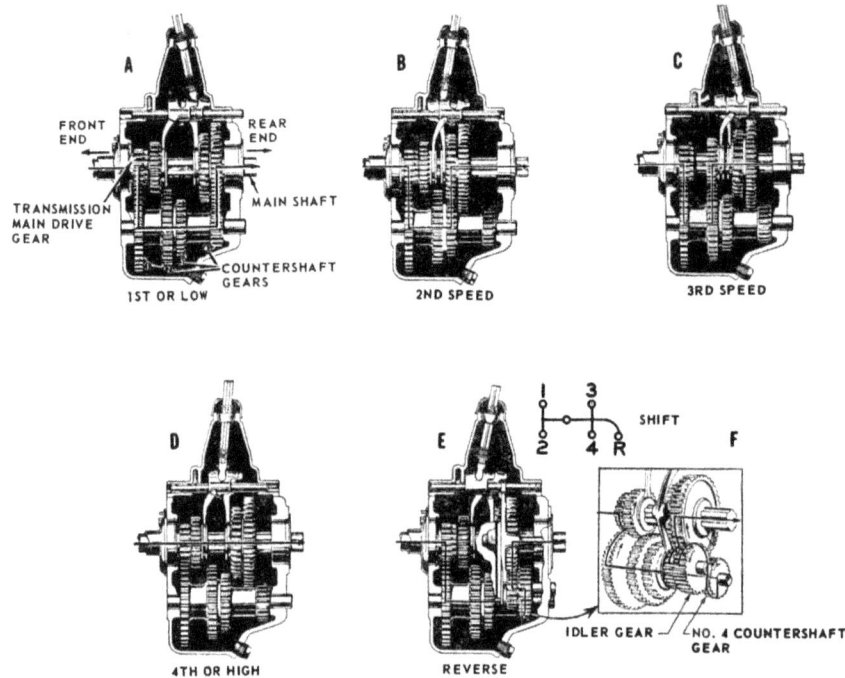

81.182

Figure 14-6.—Power flow through a 4-speed transmission.

The second speed position is obtained by moving the gear shift lever straight back from the low speed position. You will, of course, use the clutch when shifting. In B of figure 14-6 you will see that the next to the smallest countershaft gear is in mesh with the second largest sliding gear. The largest sliding gear (shift gear) has been disengaged. The flow of power has been changed as shown by the arrow. The power transmitted to the wheels in second gear (speed) is less, but the truck will move at a greater speed than it will in low gear if the engine speed is kept the same.

In shifting from the second speed to the third speed position, you move the gear shift lever through the neutral position. This is done in all selective gear transmissions. From the neutral position the driver can select the speed position required to get the power he needs. In C of figure 14-6 you will notice that the gear shift lever is in contact with the other shifting fork, and that the forward slide gear

has been meshed with the second countershaft gear. The power flow through the transmission has again been changed, as indicated by the arrow, and the truck will move at an intermediate speed between second and high.

You shift into fourth or high speed position by moving the top of the shift lever back and to the right from the neutral position. In the high speed position, the forward shift or sliding gear is engaged with the constant speed gear as shown in D of figure 14-6. The clutch shaft and the transmission shaft are now locked together and the power flow is in a straight line. In high, the truck propeller shaft revolves at the same speed as the engine crankshaft, or at a 1 to 1 ratio.

You shift to reverse by moving the top of the gear shift lever to the far right and then to the rear. Most trucks have a trigger arrangement at the gear shift ball to unlock the lever so that it can be moved from neutral to the far right. The lock prevents unintentional

135

shifts into reverse. Never attempt to shift into reverse until the forward motion of the vehicle has been completely stopped.

In F of figure 14-6, you can see how the idler gear fits into the transmission gear train. In E of figure 14-6, you can see what happens when you shift into reverse. An additional shifting fork is contacted by the shift lever in the far right position. When the shift to reverse is completed, this fork moves the idling gear into mesh with the small countershaft gear and the large sliding gear at the same time. The small arrows in the inset show how the engine power flows through the transmission to move the propeller shaft and the wheels in a reverse direction.

The different combination of gears in the transmission case makes it possible to change the vehicle speed while the engine speed remains the same. It is all a matter of gear ratios. That is, having large gears drive small gears, and small gears drive large gears. If a gear with 100 teeth drives a gear with 25 teeth, the small gear will travel four times as fast as the large one. You have stepped up the speed. Now, let the small gear drive the large gear, and the large gear will make one revolution for every four of the small gear. You have reduced speed, and the ratio of gear reduction is 4 to 1.

In the truck transmission just described, the gear reduction in low gear is 7 to 1 from the engine to the propeller shaft. In high gear the ratio is 1 to 1, and the propeller shaft turns at the same speed as the engine. This holds true for most transmissions. The second and third speed positions provide intermediate gear reductions between low and high. The gear ratio in second speed is 3.48 to 1, and in third is 1.71 to 1. The gear reduction or gear ratio in reverse is about the same as it is in low gear, and the propeller shaft makes one revolution for every seven revolutions of the engine.

All transmissions do not have four speeds forward, and the gear reductions at the various speeds are not necessarily the same. Passenger cars, for example, usually have only three forward speeds and one reverse speed. Their gear ratios are about 3 to 1 in both low and reverse gear combinations. You must remember, the gear reduction in the transmission is only between the engine and the propeller shaft. Another reduction gear ratio is provided in the rear axle assembly. If you have a common rear axle ratio of about 4 to 1, the gear reduction from the engine of a passenger car to the rear wheels in low gear would be approximately 12 to 1. In high gear the ratio would be 4 to 1 as there would be no reduction of speed in the transmission.

CONSTANT MESH TRANSMISSION

To eliminate the noise developed in the old-type spur-tooth gears used in the sliding gear transmission, the automotive manufacturers developed the constant-mesh transmission which contains helical gears.

In this type of transmission certain countershaft gears are constantly in mesh with the main shaft gears. The main shaft meshing gears are arranged so that they cannot move endwise. They are supported by roller bearings so that they can rotate independently of the main shaft (figs. 14-7 and 14-8).

In operation, when the shift lever is moved to third, the third and fourth shifter fork moves the clutch gear (A, fig. 14-8) toward the third speed gear (D, fig. 14-8). This engages the external teeth of the clutch gear with the internal teeth of the third speed gear. Since the third speed gear is rotating with the rotating countershaft gear, the clutch gear must also rotate. The clutch gear is splined to the main shaft, and therefore the main shaft rotates with the clutch gear. This principle is carried out when the shift lever moves from one speed to the next.

Constant-mesh gears are seldom used for all speeds. Common practice is to use such gears for the higher gears, with sliding gears for first and reverse speeds, or for reverse only. When the shift is made to first or reverse, the first and reverse sliding gear is moved to the left on the main shaft. The inner teeth of the sliding gear mesh with the main shaft first gear.

SYNCHROMESH TRANSMISSION

The synchromesh transmission is a type of constant-mesh transmission that permits gears to be selected without clashing, by synchronizing the speeds of mating parts before they engage. It employs a combination metal-to-metal friction cone clutch and a dog or gear positive clutch to engage the main drive gear and second-speed main shaft gear with the transmission main shaft. The friction cone clutch engages first, bringing the driving and driven members to the same speed, after which the dog clutch engages easily without clashing. This process is accomplished

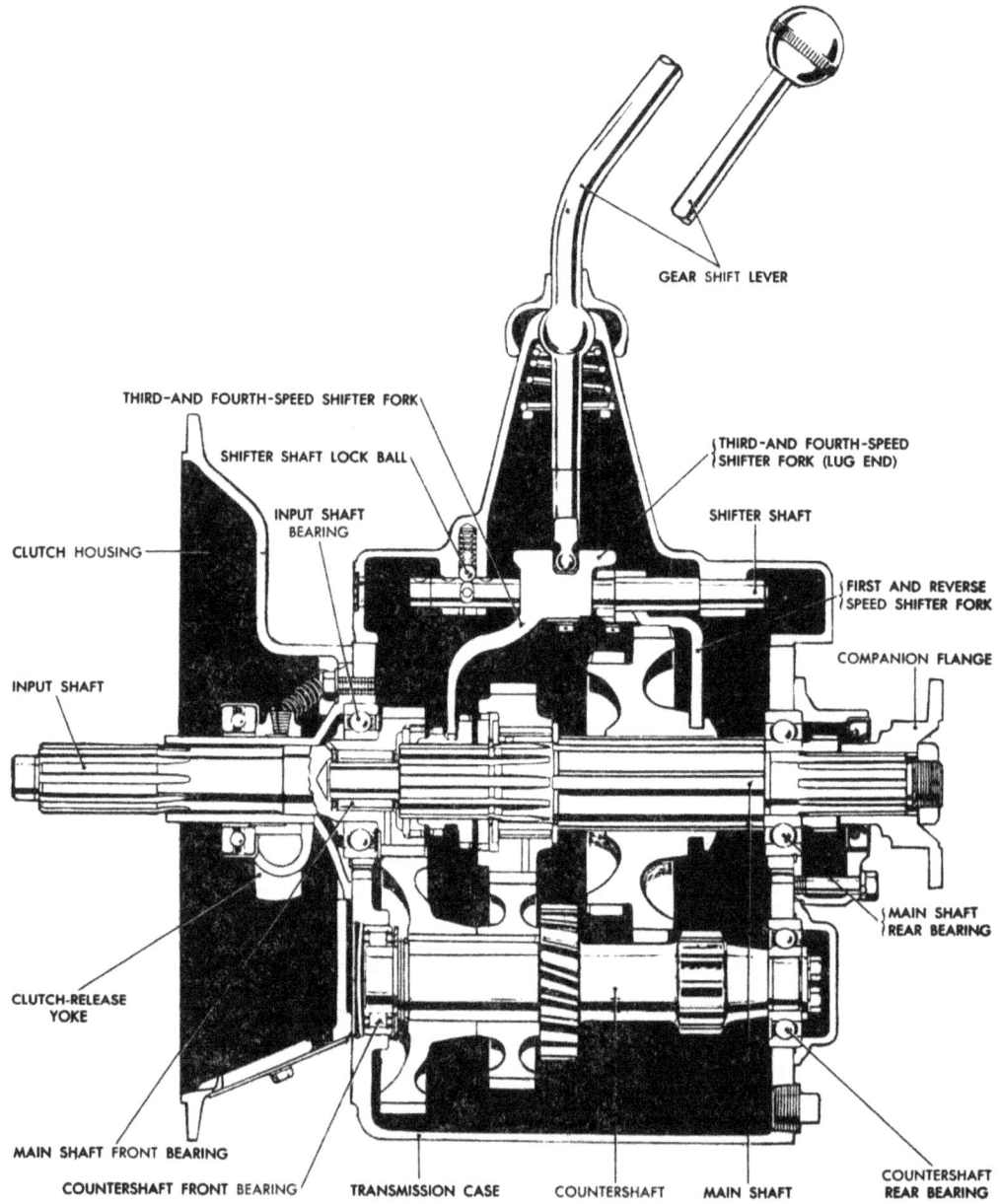

GEAR SHIFT LEVER

THIRD-AND FOURTH-SPEED SHIFTER FORK

SHIFTER SHAFT LOCK BALL

INPUT SHAFT BEARING

CLUTCH HOUSING

THIRD-AND FOURTH-SPEED SHIFTER FORK (LUG END)

SHIFTER SHAFT

FIRST AND REVERSE SPEED SHIFTER FORK

COMPANION FLANGE

INPUT SHAFT

MAIN SHAFT REAR BEARING

CLUTCH-RELEASE YOKE

MAIN SHAFT FRONT BEARING

COUNTERSHAFT FRONT BEARING

TRANSMISSION CASE

COUNTERSHAFT

MAIN SHAFT

COUNTERSHAFT REAR BEARING

81.183

Figure 14-7.—Constant-mesh transmission assembly—sectional view.

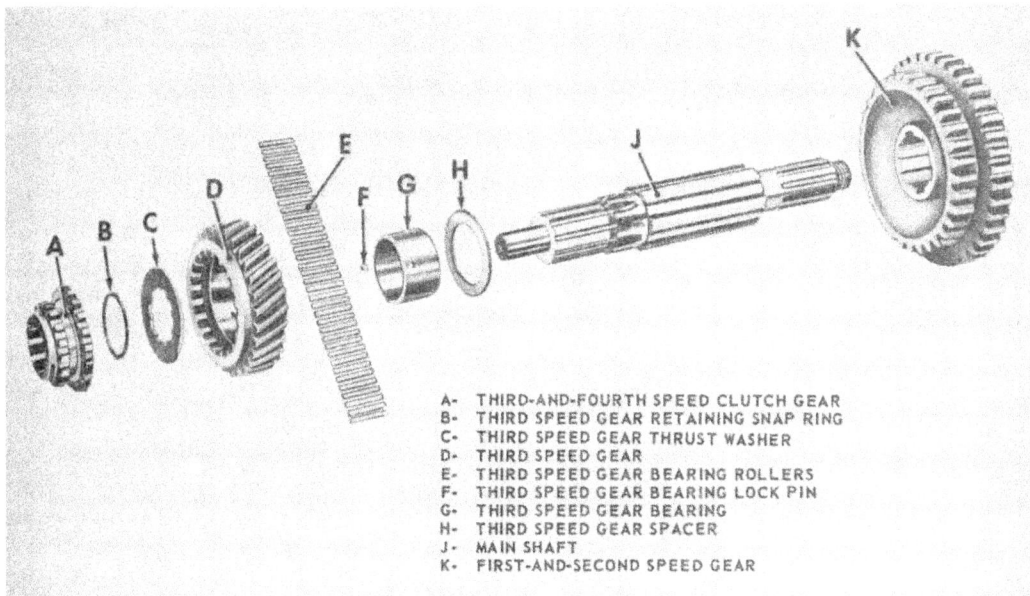

A- THIRD-AND-FOURTH SPEED CLUTCH GEAR
B- THIRD SPEED GEAR RETAINING SNAP RING
C- THIRD SPEED GEAR THRUST WASHER
D- THIRD SPEED GEAR
E- THIRD SPEED GEAR BEARING ROLLERS
F- THIRD SPEED GEAR BEARING LOCK PIN
G- THIRD SPEED GEAR BEARING
H- THIRD SPEED GEAR SPACER
J- MAIN SHAFT
K- FIRST-AND-SECOND SPEED GEAR

81.184

Figure 14-8.—Disassembled main shaft assembly.

in one continuous operation when the driver declutches and moves the control lever in the usual manner. The construction of synchromesh transmissions varies somewhat with different manufacturers, but the principle is the same in all.

The construction of a popular synchromesh clutch is shown in figure 14-9. The driving member consists of a sliding gear splined to the transmission main shaft with bronze internal cones on each side. It is surrounded by a sliding sleeve having internal teeth that are meshed with the external teeth of the sliding gear. The sliding sleeve is grooved around the outside to receive the shift fork. Six spring-loaded balls in radially-drilled holes in the gear fit into an internal groove in the sliding sleeve and prevent it from moving endwise relative to the gear until the latter has reached the end of its travel. The driven members are the main drive gear and second-speed main shaft gear, each of which has external cones and external teeth machined on its sides to engage the internal cones of the sliding gear and the internal teeth of the sliding sleeve.

The synchromesh clutch operates as follows: when the transmission control lever is moved by the driver to the third-speed or direct-drive position, the shift fork moves the sliding gear and sliding sleeve forward as a unit until the internal cone on the sliding gear engages the external cone on the main drive gear. This action brings the two gears to the same speed and stops endwise travel of the sliding gear. The sliding sleeve then slides over the balls and silently engages the external teeth on the main drive gear, locking the main drive gear and transmission main shaft together as shown in figure 14-9. When the transmission control lever is shifted to the second-speed position, the sliding gear and sleeve move rearward and the same action takes place, locking the transmission main shaft to the second-speed main shaft gear. The synchromesh clutch is not applied to first speed or to reverse. First speed is engaged by an ordinary dog clutch when constant mesh is employed, or by a sliding gear; reverse is always engaged by means of a sliding gear. Figure 14-10 shows a cross section of a synchromesh transmission which uses constant-mesh helical gears for the

138

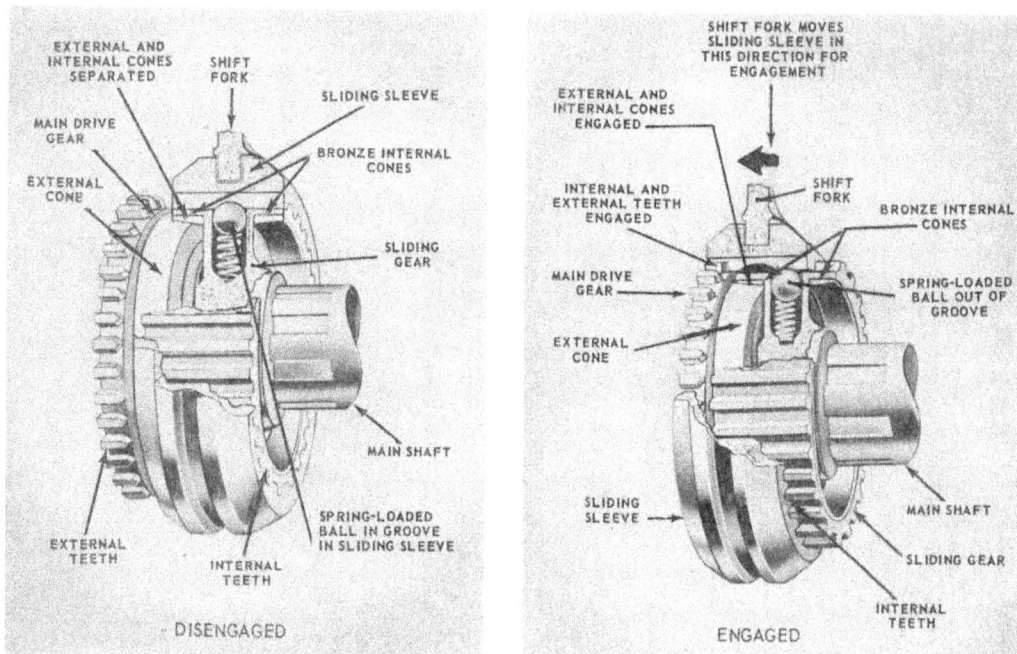

81.185

Figure 14-9.—Synchromesh clutch—disengaged and engaged.

three forward speeds and a sliding spur gear for reverse.

Some transmissions are controlled by a steering column control lever (fig. 14-11). The positions for the various speeds are the same as those for the vertical control lever except that the lever is horizontal. The shifter forks are pivoted on bellcranks which are turned by a steering column control lever through the linkage shown. The poppets shown in figure 14-10 engage notches at the inner end of each bell crank. Other types of synchromesh transmissions controlled by steering column levers have shifter shafts and forks moved by a linkage similar to those used with a vertical control lever.

AUXILIARY TRANSMISSION

The auxiliary transmission allows a rather small engine to move heavy loads in trucks by increasing the engine-to-axle gear ratios. The auxiliary transmission provides a link in the power trains of construction vehicles to divert engine power to drive 4 and 6 wheels, and also to operate accessory equipment through transfer cases and power takeoff units. (See fig. 14-12).

Trucks require a greater engine-to-axle gear ratio than passenger cars, particularly when manufacturers put the same engine in both types of equipment. In a truck, the auxiliary transmission doubles the mechanical advantage. It is connected to the rear of the main transmission by a short propeller shaft and universal joint. Its weight is supported on a frame cross-member as shown in figure 14-12. The illustration also shows how the shifting lever would extend into the driver's compartment near the lever operating the main transmission.

In appearance and in operation, auxiliary transmissions are similar to main transmissions, except that some may have two and some three speeds (low, direct and overdrive).

139

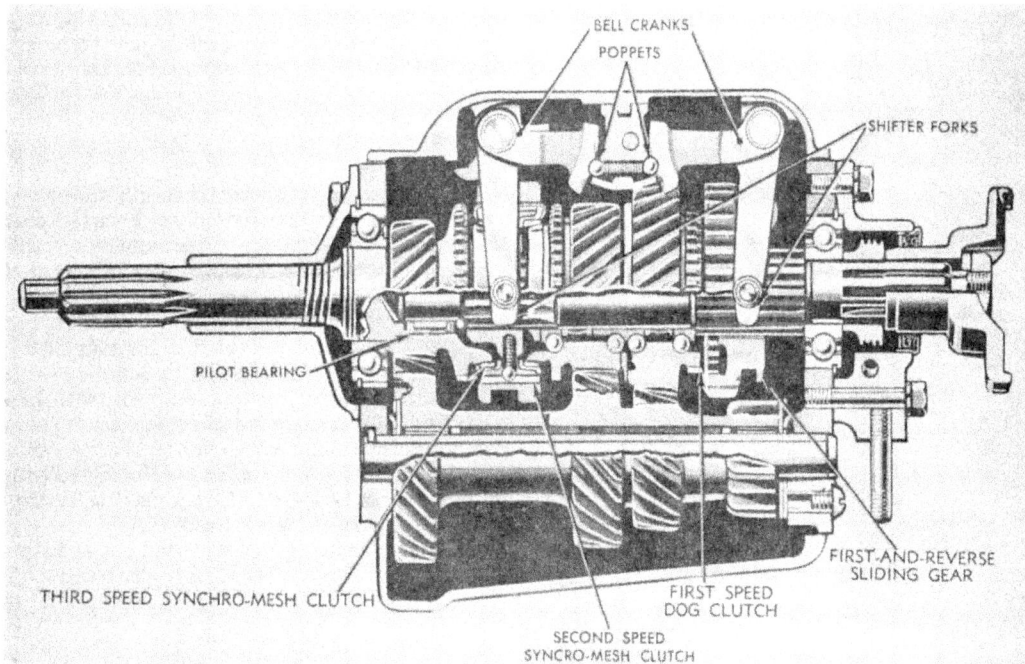

81.186

Figure 14-10.—Synchromesh transmission arranged for steering column control.

TRANSFER CASES

Transfer cases are placed in the power trains of vehicles driven by all wheels. Their purpose is to provide the necessary offsets for additional propeller shaft connections to drive the wheels.

Transfer cases in heavier vehicles have two speed positions and a declutching device for disconnecting the front driving wheels. Two speed transfer cases like the one shown in figure 14-13 serve also as auxiliary transmissions.

Some transfer cases are quite complicated. When they have speed changing gears, declutching devices, and attachments for three or more propeller shafts, they are even larger than the main transmission. A cross section of a common type of two-speed transfer case is shown in figure 14-14. Compare it with the actual installation in figure 14-13.

The declutching mechanism for the front wheels consists of a sliding sleeve spline clutch.

This same type of transfer case is used for a 6-wheel drive vehicle. The additional propeller shaft connects the drive shaft of the transfer case to the rearmost axle assembly. It is connected to the transfer case through the transmission brake drum.

Some transfer cases contain an overrunning sprag unit (or units) on the front output shaft. (A sprag unit is a form of overrunning clutch; power can be transmitted through it in one direction but not in the other.)

On these units the transfer is designed to drive the front axle slightly slower than the rear axle. During normal operation, when both front and rear wheels turn at the same speed, only the rear wheels drive the vehicle. However, if the rear wheels should lose traction and begin to slip, they tend to turn faster than the front wheels. As this happens, the sprag unit automatically engages so that the front wheels also drive the vehicle. The sprag unit simply provides an automatic means of engaging the

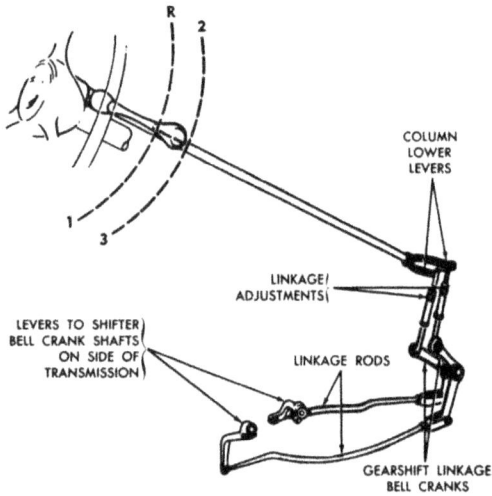

81.187

Figure 14-11.—Steering column transmission control lever and linkage.

front wheels in drive whenever additional tractive effort is required. There are two types of sprag-unit-equipped transfers, a single-sprag-unit transfer and a double-sprag-unit transfer. Essentially, both types work in the same manner.

POWER TAKEOFFS

Power takeoffs are attachments in the power train for power to drive auxiliary accessories. They are attached to the transmission, auxiliary transmission, or transfer case. A common type of power takeoff is the single-gear, single-speed type shown in figure 14-15. This unit is bolted to an opening provided in the side of the transmission case as shown in figure 14-12. The sliding gear of the power takeoff will then mesh with the transmission countershaft gear. The operator can move a shifter shaft control lever to slide the gear in and out of mesh with the counter shaft gear. The spring-loaded ball holds the shifter shaft in position.

On some vehicles you will find power takeoff units with gear arrangements that will give

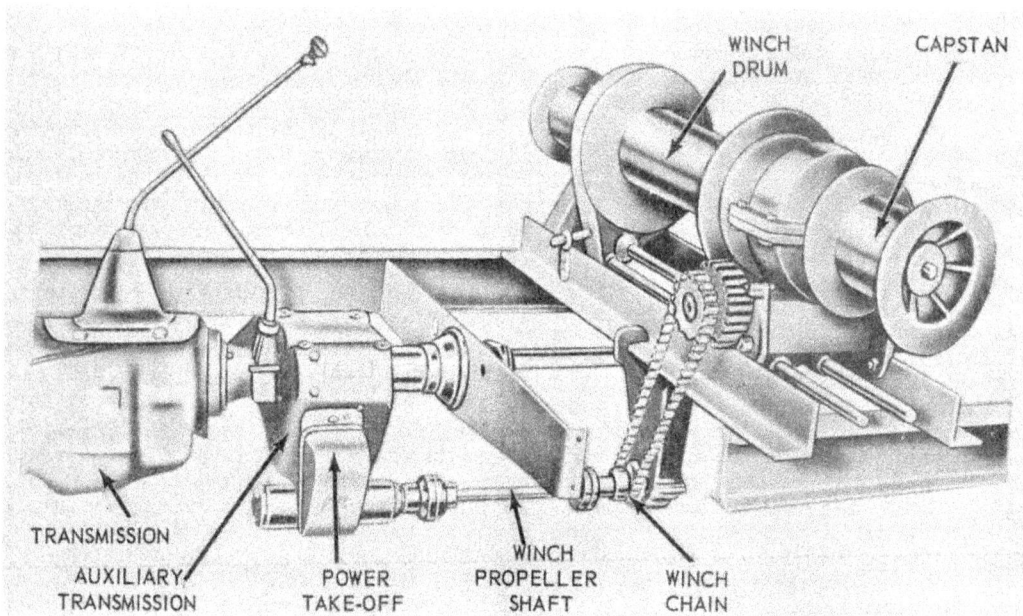

81.188

Figure 14-12.—Auxiliary transmission power takeoff driving winch.

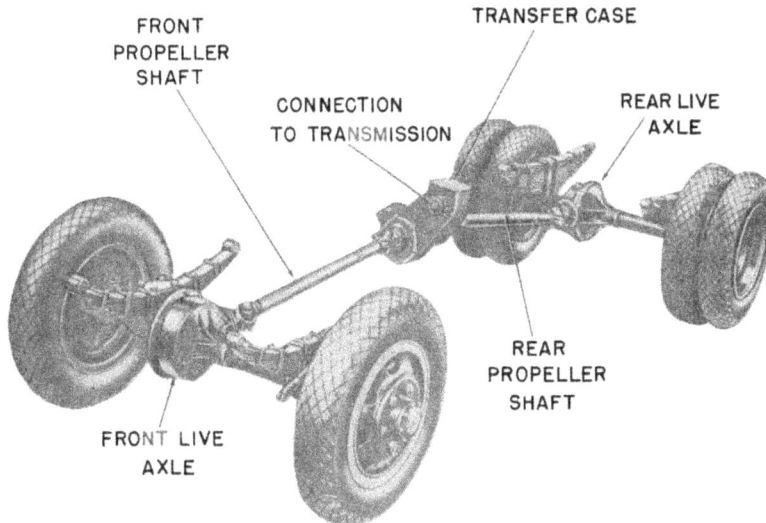

81.190

Figure 14-13.—Transfer case installed in a 4-wheel drive truck.

two speeds forward and one in reverse. Several forward speeds and a reverse gear arrangement are usually provided in power take-off units which operate winches and hoists. Their operation is about the same as the single speed units.

PROPELLER SHAFT ASSEMBLIES

The propeller shaft assembly consists of a propeller shaft, a slip joint, and one or more universal joints. This assembly provides a flexible connection through which power is transmitted from the transmission to the live axles.

The propeller shaft may be solid or tubular. A solid shaft is somewhat stronger than a hollow or tubular shaft of the same diameter, but a hollow shaft is stronger than a solid shaft of the same weight. Solid shafts are generally used inside of a shaft housing that encloses the entire propeller shaft assembly. These are called torque tube drives.

A slip joint is provided at one end of the propeller shaft to take care of end play. The driving axle, being attached to the springs, is free to move up and down while the transmission is attached to the frame and cannot move. Any upward or downward movement of the axle, as the springs are flexed, shortens or lengthens the distance between the axle assembly and the transmission. To compensate for this changing distance, the slip joint is provided at one end of the propeller shaft.

The usual type of slip joint consists of a splined stub shaft, welded to the propeller shaft, which fits into a splined sleeve in the universal joint. A cross-sectional view of the slip joint and universal joint is shown in figure 14-16.

A universal joint is a connection between two shafts that permits one to drive the other at an angle. Passenger vehicles and trucks usually have universal joints at both ends of the propeller shaft.

Universal joints are double-hinged with the pins of the hinges set at right angles. They are made in many different designs, but they all work on the same principle. (See chapter 11.)

FINAL DRIVES

A final drive is that part of the power train that transmits the power delivered through the

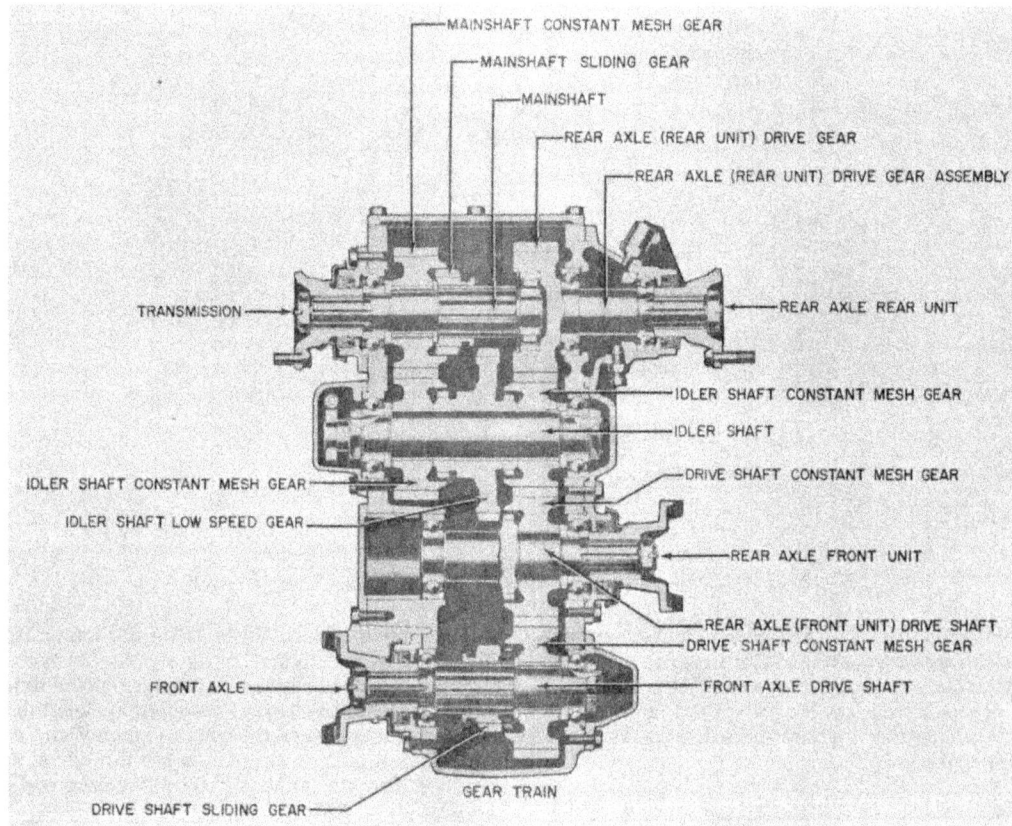

81.191

Figure 14-14.—Cross section of a 2-speed transfer case.

propeller shaft to the drive wheels or sprock-ets. Because it is encased in the rear axle housing, the final drive is usually referred to as a part of the rear axle assembly. It consists of two gears called the ring gear and pinion. These may be spur, spiral, or hypoid beveled gears, or wormgears, as illustrated in figure 14-17.

The function of the final drive is to change by 90 degrees the direction of the power trans-mitted through the propeller shaft to the driving axles. It also provides a fixed reduction be-tween the speed of the propeller shaft and the axle shafts and wheels. In passenger cars this

reduction varies from about 3 to 1 to 5 to 1. In trucks, it can vary from 5 to 1 as much as 11 to 1.

The gear ratio of a final drive having bevel gears is found by dividing the number of teeth on the drive gear by the number of teeth on the pinion. In a worm gear final drive, the gear ratio is found by dividing the number of teeth on the gear by the number of threads on the worm.

Most final drives are of the gear type. Hy-poid gears are used in passenger cars and light trucks to give more body clearance. They per-mit the bevel drive pinion to be placed below the center of the bevel drive gear, thereby

143

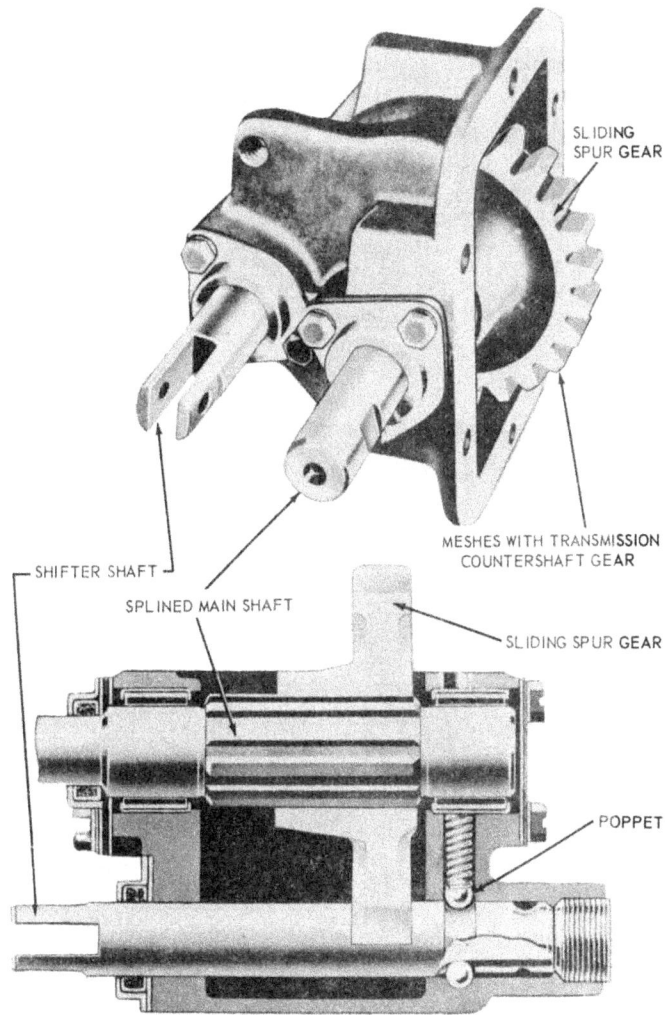

81.192

Figure 14-15.—Single speed, single gear, power takeoff.

lowering the propeller shaft (see fig. 14-17). Worm gears allow a large speed reduction and are used extensively in the larger trucks. Spiral bevel gears are similar to hypoid gears. They are used in both passenger cars and trucks to replace spur gears that are considered too noisy.

DIFFERENTIALS

The construction and principles of operation of the gear differential were described in chapter 11 of this book. We will briefly review some of the high points of that chapter here, and then go on to describe some of the more common types

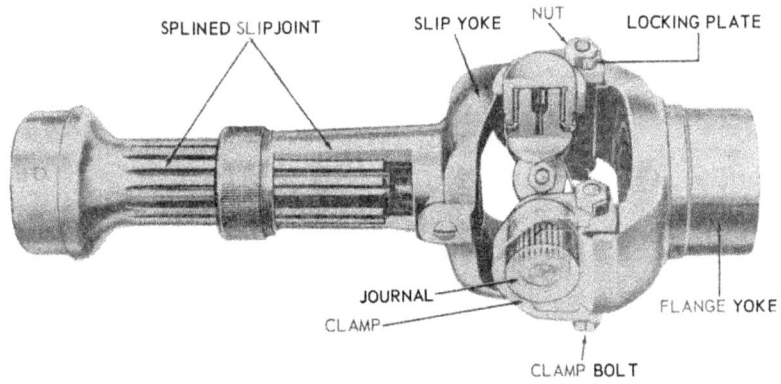

SPLINED SLIPJOINT   SLIP YOKE   NUT   LOCKING PLATE

JOURNAL

CLAMP

CLAMP BOLT

FLANGE YOKE

2.200

Figure 14-16.—Slip joint and common type of universal joint.

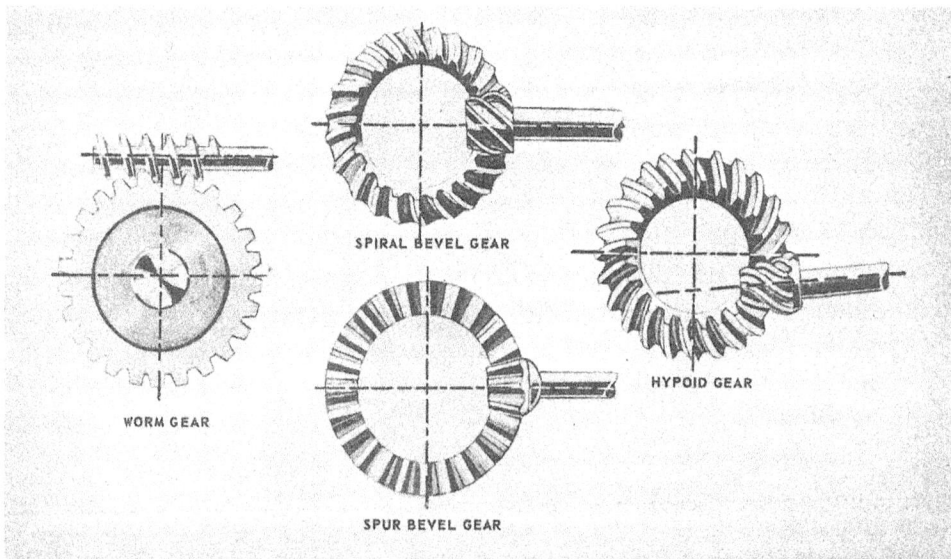

WORM GEAR

SPIRAL BEVEL GEAR

HYPOID GEAR

SPUR BEVEL GEAR

81.195

Figure 14-17.—Gears used in final drives.

145

of gear differentials as applied in automobiles and trucks.

The purpose of the differential is easy to understand when you compare a vehicle to a company of men marching in mass formation. When the company makes a turn, the men in the inside file must take short steps, almost marking time, while men in the outside file must take long steps and walk a greater distance to make the turn. When a motor vehicle turns a corner, the wheels on the outside of the turn must rotate faster and travel a greater distance than the wheels on the inside. This causes no difficulty for front wheels of the usual passenger car because each wheel rotates independently on opposite ends of a dead axle. However, in order to drive the rear wheels at different speeds, the differential is needed. It connects the individual axle shaft for each wheel to the bevel drive gear. Therefore, each shaft can turn at a different speed and still be driven as a single unit. Refer to the illustration in figure 14-18 as you study the following discussion on differential operation.

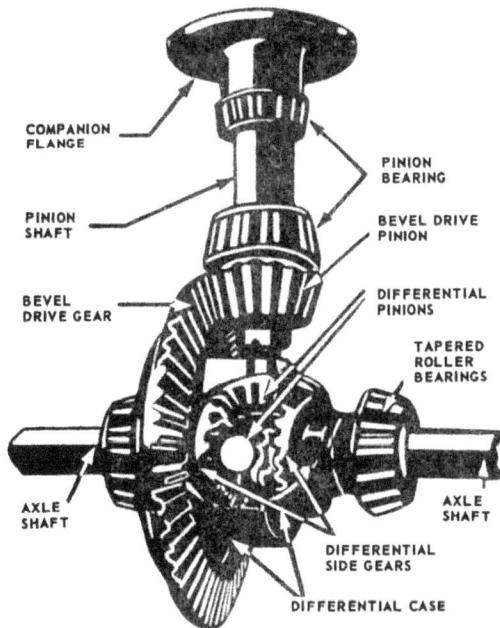

COMPANION FLANGE

PINION BEARING

PINION SHAFT

BEVEL DRIVE PINION

BEVEL DRIVE GEAR

DIFFERENTIAL PINIONS

TAPERED ROLLER BEARINGS

AXLE SHAFT

AXLE SHAFT

DIFFERENTIAL SIDE GEARS

DIFFERENTIAL CASE

81.196
Figure 14-18.—Differential with part of case cut away.

The differential described in chapter 11 had two inputs and a single output. The differential as used in the automobile, however, has a single input and two outputs, the input being introduced from the propeller shaft, and the outputs going to the rear axles and wheels. In this discussion, the "spider gears" are referred to as "differential pinions," so don't let this confuse you.

The bevel drive pinion, connected to the propeller shaft, drives the bevel drive gear and the differential case to which it is attached. Therefore, the entire differential case always rotates with the bevel drive gear whenever the propeller shaft is transmitting rotary motion. Within the case, the differential pinions are free to rotate on individual shafts called trunnions. These trunnions are attached to the walls of the differential case, so that whenever the case is turning, the differential pinions must revolve—one about the other—in the same plane as the bevel drive gear.

The differential pinions mesh with the side gears, as did the spider and side gears in the differential described in chapter 11. The axle shafts are splined to the side gears and keyed to the wheels. Power is transmitted to the axle shafts through the differential pinions and the side gears. When resistance is equal on each rear wheel, the differential pinions, side gears, and axle shafts all rotate as one unit with the bevel drive gear. In this case, there is no relative motion between the pinions and the side gears in the differential case. That is, the pinions do not turn on the trunnions, and their teeth will not move over the teeth of the side gears.

When the vehicle turns a corner, one wheel must turn faster than the other. The side gear driving the outside wheel will run faster than the side gear connected to the axle shaft of the inside wheel. To compensate for this difference in speed, and to remain in mesh with the two side gears, the differential pinions must then turn on the trunnions. The average speed of the two side gears, axle shafts, or wheels is always equal to the speed of the bevel drive gear.

To overcome the situation where one spinning wheel might be undesirable, some trucks are provided with a differential lock. This is a simple dog clutch, controlled manually or automatically, which locks one axle shaft to the differential case and bevel drive gear. Although this device forms a rigid connection between the two axle shafts and makes both wheels rotate at

146

the same speed, it is used very little. Too often the driver forgets to disengage the lock after using it. There are, however, automatic devices for doing almost the same thing. One of these, which is rather extensively used today, is the high-traction differential. It consists of a set of differential pinions and side gears which have fewer teeth and a different tooth form from the conventional gears. Figure 14-19 shows a comparison between these and standard gears. These differential pinions and side gears depend on a variable radius from the center of the differential pinion to the point where it comes in contact with the side gear teeth, which is, in effect, a variable lever arm. As long as there is relative motion between the pinions and side gears, the torque is unevenly divided between the two driving shafts and wheels; whereas, with the usual differential, the torque is evenly divided at all times. With the high-traction differential, the torque becomes greater on one wheel and less on the other as the pinions move around, until both wheels start to rotate at the same speed. When this occurs, the relative motion between the pinion and side gears stops and the

torque on each wheel is again equal. This device assists considerably in starting the vehicle or keeping it rolling in cases where one wheel encounters a slippery spot and loses traction while the other wheel is on a firm spot and has traction. It will not work, however, when one wheel loses traction completely. In this respect it is inferior to the differential lock.

With the no-spin differential (fig. 14-20). one wheel cannot spin because of loss of tractive effort and thereby deprive the other wheel of driving effort. For example, one wheel is on ice and the other wheel is on dry pavement. The wheel on ice is assumed to have no traction. However, the wheel on dry pavement will pull to the limit of its tractional resistance at the pavement. The wheel on ice cannot spin because wheel speed is governed by the speed of the wheel applying tractive effort.

The no-spin differential does not contain pinion gears and side gears as does the conventional differential. Instead, it consists essentially of a spider attached to the differential drive ring gear through four trunnions, plus two driven clutch members with side teeth that are

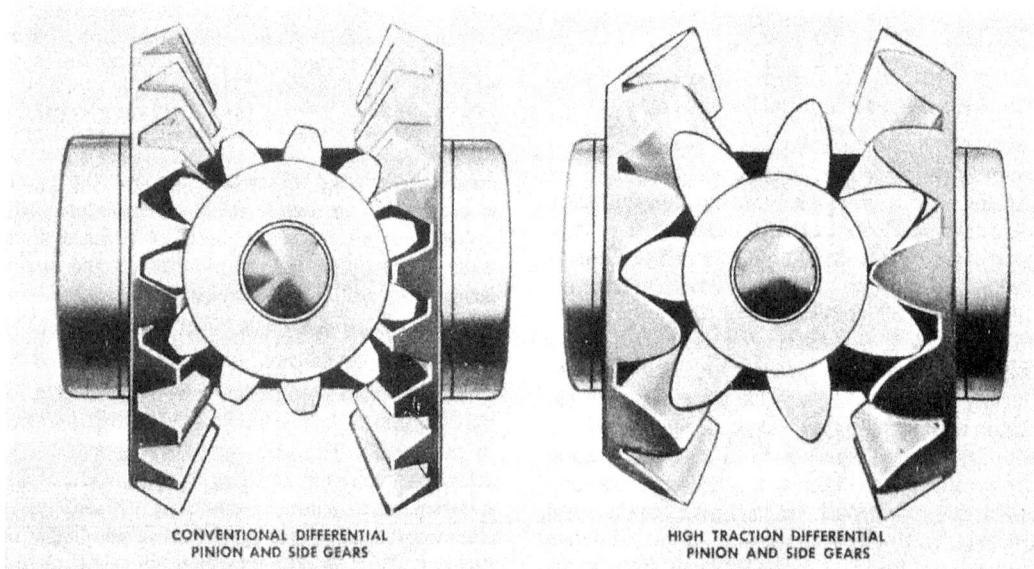

CONVENTIONAL DIFFERENTIAL
PINION AND SIDE GEARS

HIGH TRACTION DIFFERENTIAL
PINION AND SIDE GEARS

81.197

Figure 14-19.—Comparison of high-traction differential gears and standard differential gears.

147

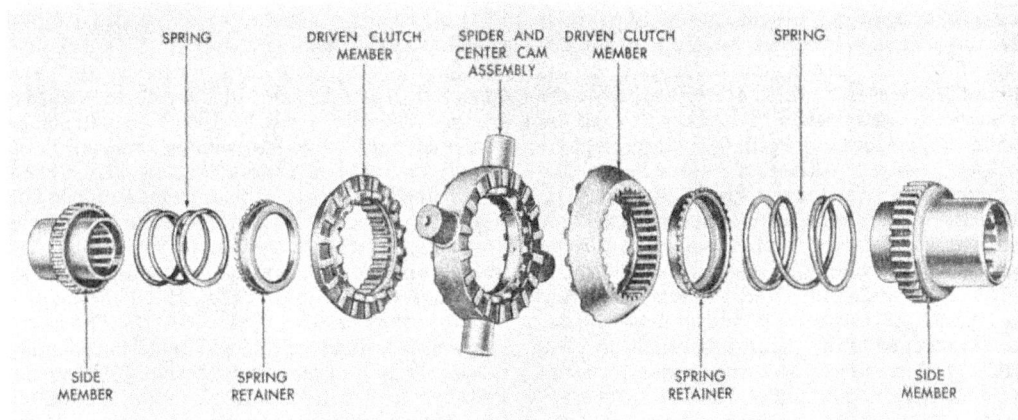

81.198

Figure 14-20.—No spin differential—exploded view.

indexed by spring pressure with side teeth in the spider. Two side members are splined to the wheel axles and in turn are splined into the driven clutch members.

AXLES

A live axle is one that supports part of the weight of a vehicle and also drives the wheels connected to it. A dead axle is one that carries part of the weight of a vehicle but does not drive the wheels. The wheels rotate on the ends of the dead axle.

Usually, the front axle of a passenger car is a dead axle and the rear axle is a live axle. In 4-wheel drive vehicles, both front and gear axles are live axles, and in 6-wheel drive vehicles, all three axles are live axles. The third axle, part of a bogie drive, is joined to the rearmost axle by a trunnion axle. The trunnion axle is attached rigidly to the frame. Its purpose is to help in distributing the load on the rear of the vehicle to the two live axles which it connects.

There are four types of live axles used in automotive and construction equipment. They are: plain, semifloating, three-quarter floating, and full floating.

The plain live axle, or nonfloating rear axle, is seldom used in equipment today. The axle shafts in this assembly are called nonfloating because they are supported directly in bearings located in the center and ends of the axle housing. In addition to turning the wheels, these shafts carry the entire load of the vehicle on their outer ends. Plain axles also support the weight of the differential case.

The semifloating axle (fig. 14-21) that is used on most passenger cars and light trucks

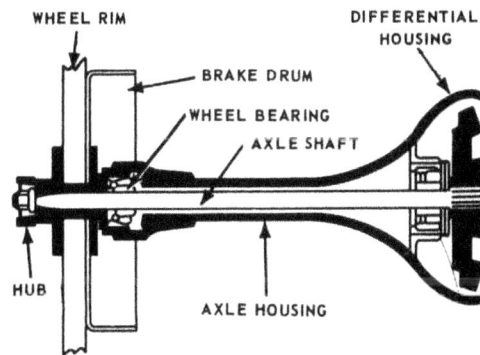

81.200

Figure 14-21.—Semifloating rear axle.

148

has its differential case independently supported. The differential carrier relieves the axle shafts from the weight of the differential assembly and the stresses caused by its operation. For this reason the inner ends of the axle shafts are said to be floated. The wheels are keyed to outer ends of axle shafts and the outer bearings are between the shafts and the housing. The axle shafts therefore must take the stresses caused by turning, skidding, or wobbling of the wheels. The axle shaft in a semifloating live axle can be removed after the wheel has been pulled off.

The axle shafts in a three-quarter floating axle (fig. 14-22) may be removed with the wheels, which are keyed to the tapered outer ends of the shafts. The inner ends of the shafts are carried as in a semifloating axle. The axle housing, instead of the shafts, carries the weight of the vehicle because the wheels are supported by bearings on the outer ends of the housing. However, axle shafts must take the stresses caused by the turning, skidding, and wobbling of the wheels. Three-quarter floating axles are used in some trucks but in very few passenger cars. The full floating axle is used in most heavy trucks. (See fig. 14-23). These axle shafts may be removed and replaced without removing the wheels or disturbing the differential. Each wheel is carried on the end of the axle tube on two ball bearings or roller bearings and the axle shafts are not rigidly connected to the wheels. The wheels are driven through a clutch arrangement or flange on the ends of the axle shaft which is bolted to the outside of the wheel hub. The bolted connection between axle and wheel does not make this assembly a true full floating axle, but nevertheless, it is called a floating axle. A true full floating axle transmits only turning effort, or torque.

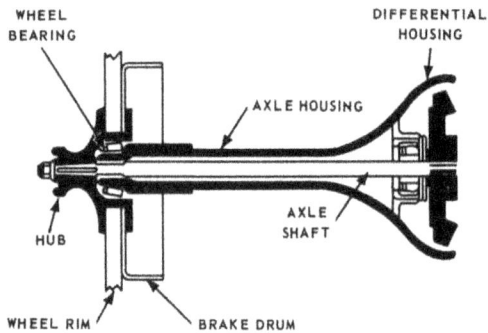

81.201

Figure 14-22.—Three-quarter floating rear axle.

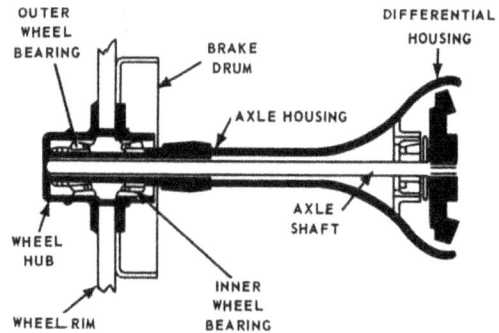

81.202

Figure 14-23.—Full floating rear axle.

CHAPTER 15

BASIC COMPUTER MECHANISMS

We have already studied several examples of complex machines in the preceding chapters to learn how simple machines and basic mechanisms are utilized in their design. The analog computer, of the kind used in modern fire control systems, is a complex machine in every sense of the word. We will not attempt in this book to break down and analyze a complete computer. We will, however, examine a few of the special devices commonly used in computers. These devices have come to be known as basic computer mechanisms. They are, however, quite complex machines in themselves—as you'll soon agree. Like the engine, the typewriter, and the other machines we've studied, these mechanisms are only combinations of simple machines cleverly designed to do a specific kind of work. As before, the watchword is Look For the Simple Machines.

DIFFERENTIALS

The differentials used in the analog computer are gear differentials similar to those described in chapter 11. They are different from the automobile differential in that instead of receiving a single input and delivering two outputs, they receive two inputs and combine them into a single output. Most of the differentials in a computer are quite small, averaging about 2'' x 2 1/2'' in size, and are designed for light loads. Some computers may have as many as 150 gear differentials in their makeup.

Figure 15-1 illustrates the symbol used to indicate the differential in schematic drawings. The cross in the center represents the spider. The arrows pointing inward represent inputs, and the arrow pointing outward is the output.

Figure 15-2 shows one of the many applications of the gear differential in a computer. In this case, the differential is being used as an integral part of a followup control. Computing mechanisms are not designed to drive heavy loads. The outputs from such mechanisms often merely control the action of servomotors. The motors do the actual driving of the loads to be handled. The device which makes it possible for the comparatively weak output from a computing mechanism to control the action of a servomotor is called a followup control. In this device, the differential is used to measure the difference, or "error," in position between the input and the output. The input is geared to one side of the differential. The servo output is used to do two things: (1) to position whatever mechanism is being handled, and (2) to drive the other side of the differential. This second operation is known as the servo "response."

When there is a difference between the input and the output, the spider of the differential turns. As this happens, the spider shaft operates a set of controls which control the action of the servomotor in such a way that the motor drives its side of the differential in a direction opposite to that taken by the input. That is, the servo always drives to reduce the difference, or error, to zero.

LINKAGES FOR ADDING AND SUBTRACTING

Addition of two quantities is performed in the linkage mechanism by means of adding levers as shown in figure 15-3. In the example two quantities, designated X and Y, are to be added. Their values are represented by the movements of the two slide bars. The adding lever is pivoted at its center to another slide bar, and its opposite ends are connected through links to the X and Y slides. To illustrate the problem, scales showing the values of the quantities represented by movements of slides have been drawn in the figure, and index marks are placed on the slides. The units on the center scale are half as large as those on the other two.

12.87

Figure 15-1.—This is the symbol used to indicate the differential in schematic drawings.

If the Y slide is held in place and the X slide is moved, the adding lever pivots about its lower end. The center slide, which is connected to the midpoint of the lever then moves half as far as the X slide. If the X movement is one unit, the center slide also moves one unit since the units on the center scale are half as large as those on the X scale. Simiarly, movements of the Y slide with the X slide held in place add one unit on the center scale for each unit movement of

Y. At the left of figure 15-3, the parts are shown in zero position, with the three index marks opposite the zero points of the scales. At the right, the X slide has been moved one unit, and the Y slide has been moved three.

The center slide has traveled one unit in response to the X-travel and three more in response to the Y-travel, and so stands at a reading of four. Similarly, for any position of the X and Y slides, the reading on the center scale represents the quantity X plus Y.

There are several variations of the adding lever used in computing linkage, but their operating principles are the same.

MECHANICAL MULTIPLIERS

There are two basic types of mechanical multipliers—those using rotary gearing and those using linkages.

The rotary gearing type produces a solution through the use of similar triangles. There are four types of rotary multipliers in use—screw, rack, sector, and cam. Since they all operate in fundamentally the same manner, we will discuss the screw type multiplier and then compare the other types to it.

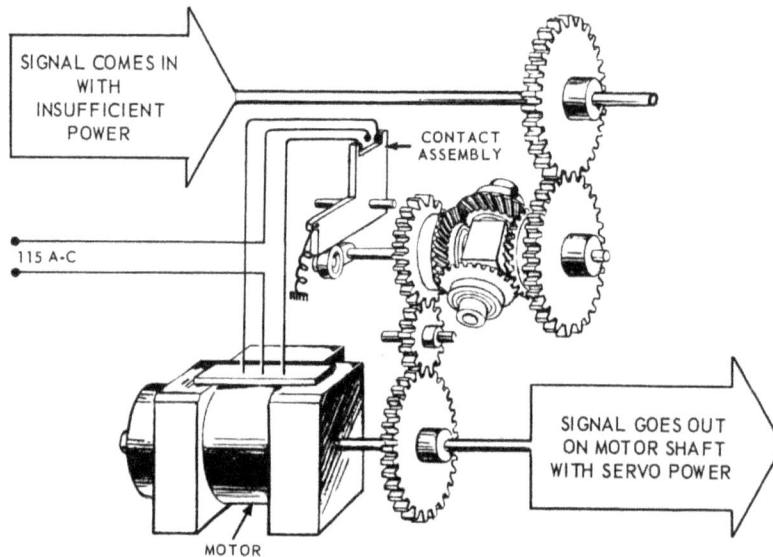

110.9

Figure 15-2.—Simplified sketch of a followup control showing application of a gear differential.

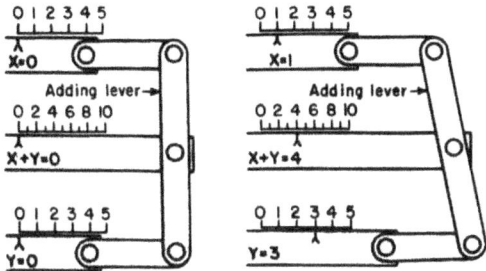

12.88

Figure 15-3.—Adding lever.

The screw multiplier, shown in figure 15-4, has two inputs and one output. The inputs are shaft values which position the input slide and input rack. The output appears at the output rack which positions the output shaft. Thus the output shaft value is always proportional to the product of the two inputs.

One input gearing is connected to two long screws. These screws pass through the threaded sleeve-like ends of the slotted input slide. As the input gears to the screws are rotated, the two screws turn to move the slide to the left or right. At the same time, the other input moves the input rack up or down, moving the slotted pivot arm around the stationary pin.

The multiplier pin is mounted in the slots of the input slide, pivot arms, and output rack, connecting all three where the slots cross. As the multiplier pin moves the input slide and pivot arm, it positions the output rack and gear.

Now, consider the multiplier in the zero position shown in figure 15-5. If only the screws are rotated, the input slide moves to the right; but it will not affect the output rack. Similarly, if only the input rack is moved up or down from the zero position, the output rack will be un-affected. This is a reasonable result, for any number multiplied by zero is equal to zero.

From this we can conclude that both inputs must be removed from the zero position for an output. Such a condition is shown in figure 15-6. Notice the triangle superimposed on the device. The value a represents the amount of rack input. The value b represents the amount of slide input. K is a fixed distance, since the multiplier pin cannot move and the input rack travels in a machined guide.

Because the angles are equal, the triangles are similar. Thus the value of X can be determined if the other values are known.

$$\left(\text{Actually, } X = \frac{ba}{K}\right)$$

This equation shows that the output (X) is always proportional to the product of the two inputs. The constant value (K) can be compensated for by the proper choice of input and output gearing for the multiplier. These

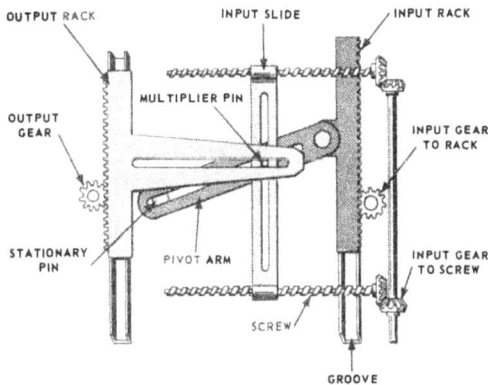

12.91

Figure 15-4.—Screw type multiplier.

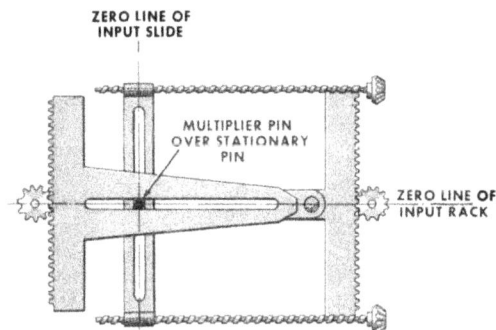

12.92

Figure 15-5.—Screw type multiplier—zero position.

152

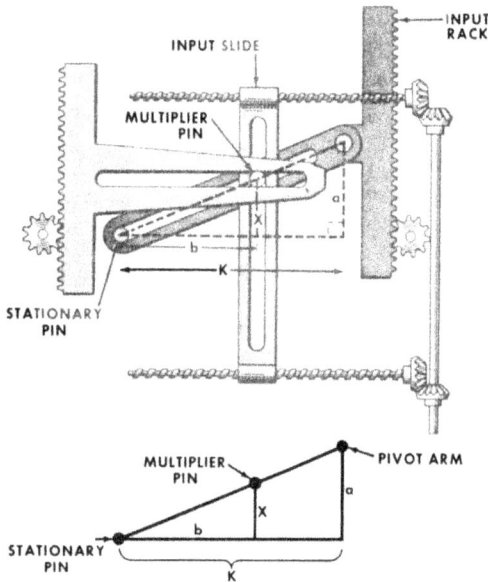

12.93

Figure 15-6.—Screw type multiplier—
multiplying positive values.

multipliers can also determine the product of negative values.

The rack type multiplier in figure 15-7 performs the same task as the screw type multiplier. The differences are that (1) the screw input has been replaced with an input rack, and (2) the output rack is placed on the same side as the second input rack.

The sector type, although different in construction, also employes triangles for the multiplication of the two inputs. A sector type multiplier is shown in figure 15-8. One input positions the input sector arm and the other input turns a large screw that is mounted on the input sector arm. The bevel gear turns this lead screw through a universal joint. The use of the universal joint permits the input to drive the lead screw as the sector arm changes its angular position. Notice that the lead screw drives the multiplier pin up and down the sector arm. Thus the position of the input sector arm and the position of the multiplier on the lead screw represents the two values to be multiplied.

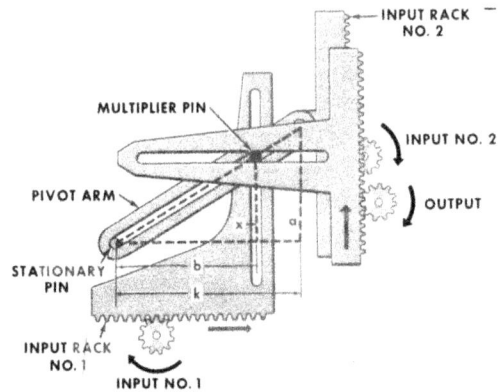

12.94

Figure 15-7.—Rack type multiplier.

A study of figure 15-9 along with figure 15-8 will reveal how the triangles are established. While the sector type multiplier can handle both positive and negative inputs on the input sector arm, the input to the lead screw must always be a positive quantity.

The cam computing multiplier is a dual operation device. It computes a function of one value on a cam and multiplies that function by a second value. It is a combination of a cam and a rack type multiplier.

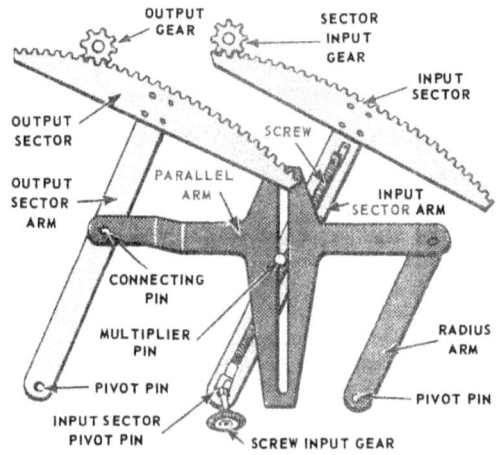

12.95

Figure 15-8.—Sector type multiplier.

153

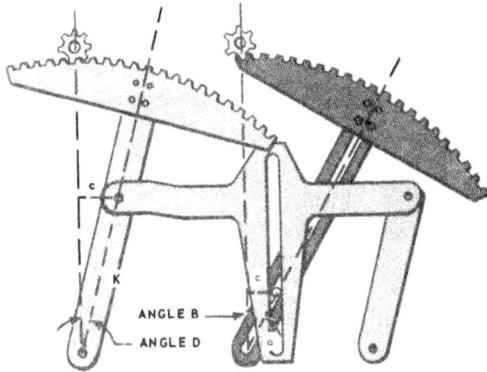

12.96

Figure 15-9.—Multiplying with the sector
type multiplier.

A single cam computing multiplier is shown in
figure 15-10. Notice it is like the rack type
multiplier except that one of the inputs is
positioned by a cam instead of a rack. The cam
follower pin is mounted directly on the multiplier input slide. This cam may be cut to
compute any desired function of the cam input.

One input drives the input rack through the
rack input gear. The other input drives the cam

directly. The cam positions the input slide
according to the function for which the cam
was cut. Thus the cam output becomes the
slide input. The position of the output rack
represents a value which is proportional to
the product of the cam output and the rack
input, just as in the rack type multiplier.

A two-cam computing multiplier computes
the function of both inputs and multiplies these
functions together. The output is proportional to
the product of the functions of the two inputs.

Linkage multipliers of the type described
here are used when one of the factors is a constant, as shown in figure 15-11. In this example
we want to transform a movement representing
the quantity X into one representing 1.5X. One
end of the multiplying lever is pivoted on the fixed
frame of the computer, as indicated by the cross-
hatched circle in the figure.

The input and output links are connected to
the lever at different points, the connection of
the output link being 1.5 times as far from the
fixed pivot as the connection of the input link.
The two scales shown in figure 15-11 have units
of the same size; but because of the difference
in lever arms, each one-unit movement of the
input link moves the output link a unit and a half.
Then, if the input movement represents the
quantity X, the output represents 1.5X.

In many cases the computing problem requires the multiplication of two variable quantities. The multiplying levers shown in figure
15-11 cannot be used for this purpose. Figure
15-12 shows a linkage designed to multiply two
variables, X and Y. The levers AB and ED
are pivoted on the fixed structure and are
connected by links BC and CD, both of which
have exactly the same length as AB. The X
input is applied by a link connected at B. The
Y input is applied by a link connected at C;

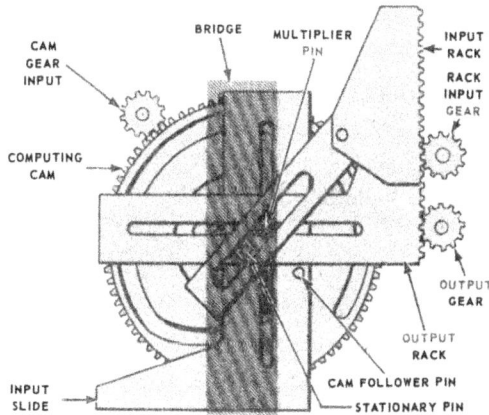

12.97

Figure 15-10.—Single cam computing multiplier.

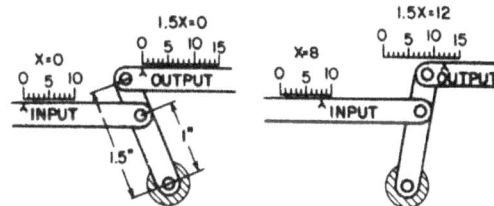

12.98

Figure 15-11.—Multiplying lever.

154

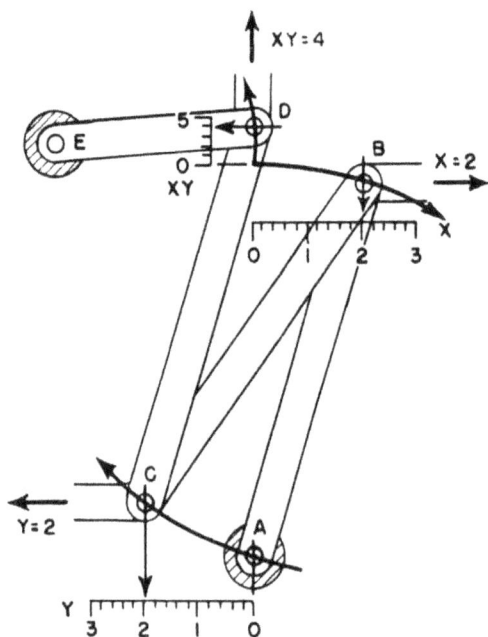

12.99

Figure 15-12.—Multiplying linkage.

and the output XY is taken off at D. In the position of the links shown in figure 15-12, X equals two units, Y equals two units, and XY equals four.

This type of linkage in the computer can be operated in reverse to serve as a divider instead of a multiplier. Two inputs are applied at the points corresponding to B and D in figure 15-12. The output taken off at C then represents the input at D divided by the input at B.

Component solvers are devices that are used in mechanical computers. The component solver takes a vector of a given magnitude and angular position, and resolves it into its two rectangular components.

There are several types of component solvers. However, this discussion will be limited to the screw type component solver. The device consists of a vector gear, two racks, two output gears, two input gears, and a screw and pin assembly. The speed input gear drives a gear train which causes the screw to turn. As the screw turns, it drives an assembly carrying the pin,

thus changing the length of the vector which is proportional to the input component, such as r (slant range). (See figure 15-13.)

In this type of component solver the pin can travel nearly the full width of the vector gear. From the pin's zero or center position it can be moved in either of two directions, which correspond to positive or negative.

The vector input gear drives the vector gear in the desired angular direction indicating target position. The pin positions the racks as it moves along the slot, thus resolving the vector into its components.

An angle resolver is a linkage mechanism which computes the sine and cosine of an angle. Figure 15-14A shows an angle resolver consisting of gear H with two crank pins M and N, mounted 90° apart and equidistant from the center of the gear. Attached to each crank pin is an output link which transmits the horizontal component of motion of the pin as the gear rotates. The horizontal component of the displacement of pin M is proportional to the sine of the angle through which gear H rotates. The horizontal component of pin N is proportional to the cosine of the angle through which gear H rotates.

Figure 15-14B shows the resolver in its zero position, with the radius OM perpendicular to the horizontal center line and the radius ON in the horizontal center line. Notice that link R (sine output) is at zero horizontal displacement, and that link S (cosine output) is at maximum

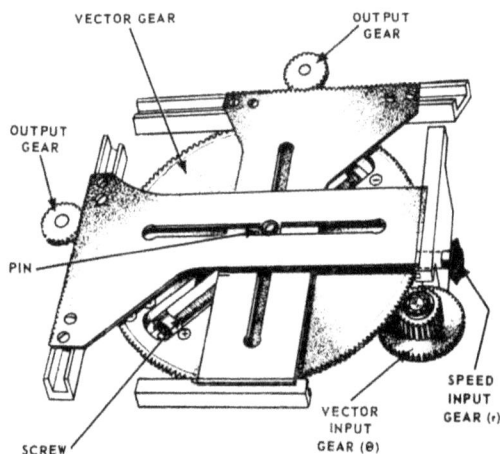

12.113

Figure 15-13.—Screw type component solver.

155

horizontal displacement. If we rotated gear H clockwise through an angle 30°, link R would move to the right and link S to the left along the horizontal center line. The linear displacement of the output links would be proportional to the sine and cosine function of the angle.

The outputs of the resolver of figure 15-14 are only approximate values. This is because the output links are not parallel to the horizontal center line. The output links have a slight angular movement that must be compensated for to eliminate distortion. This is accomplished by additional gearing, and by making the pins M and N eccentric.

Integrators, as used in computers, perform a special type of multiplication. In the disc-type integrator, illustrated in figure 15-15, a constantly changing value, such as time, is multiplied by a variable such as range rate, such as range (the rate that a target range is opening or closing), the output being a continuous value of their product which can be accumulated as shaft rotation.

The instrument consists of a flat circular disc revolved at constant speed by a motor equipped with a clock escapement; a carriage, containing two balls driven by friction with the surface of the disc, and themselves driving an output roller; and suitable shafts and gears for transmission of values to and from the unit. Rotation of the disc rotates the lower ball, which turns the upper ball, and this in turn

rotates the output roller. The balls are supported in a movable carriage so that the point of contact between the lower ball and the disc can be shifted along a diameter from the center of the disc to either edge. Spring tension on the roller provides sufficient pressure to prevent slipping. Two balls are used to reduce the sliding friction that results when only one is used.

The speed of roller rotation depends upon the speed at which the balls rotate. If the carriage is in the center of the disc, no motion is imparted to the roller. As the carriage is moved off center, the balls will begin to rotate, and will reach their maximum speed at the edge of the disc. The speed varies with the distance of the carriage from the center. Values of rotation on one side of center are considered positive, while if the carriage is moved to the opposite side, rotation will be in the opposite direction and will give negative output values.

SUMMARY

Of the many existing complex machines to choose from, you have been given only a few examples to study. The operational principles of some of them may have come to you quite easily—others may have been a bit harder to grasp. In any case, if you'll keep firmly in mind the following points that have been brought out in this book—you'll find all machines much easier to analyze and understand.

A machine is any device that helps you do work. It helps you by changing motion, magnitude, or speed of the effort you apply.

All machines consist of one or more of the six basic, or simple machines. These are the lever, the block and tackle, the wheel and axle, the inclined plane, the screw, and gears.

When machines give a mechanical advantage of more than one, they multiply the force of your effort. When they give a mechanical advantage of less than one, they multiply either the motion, or the speed of the force you apply.

No machine is 100 percent efficient. Some of your effort is always used to overcome friction. You always do more work on the machine than it does on the load.

You can figure out how any complex machine works by breaking it down into the simple machines from which it is made, and following the action through, step by step.

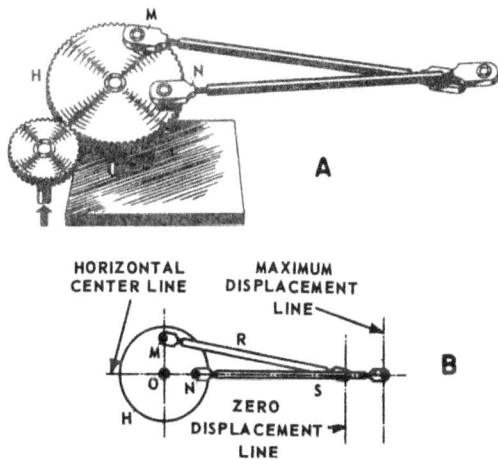

12.115
Figure 15-14.—Basic linkage angle resolver.

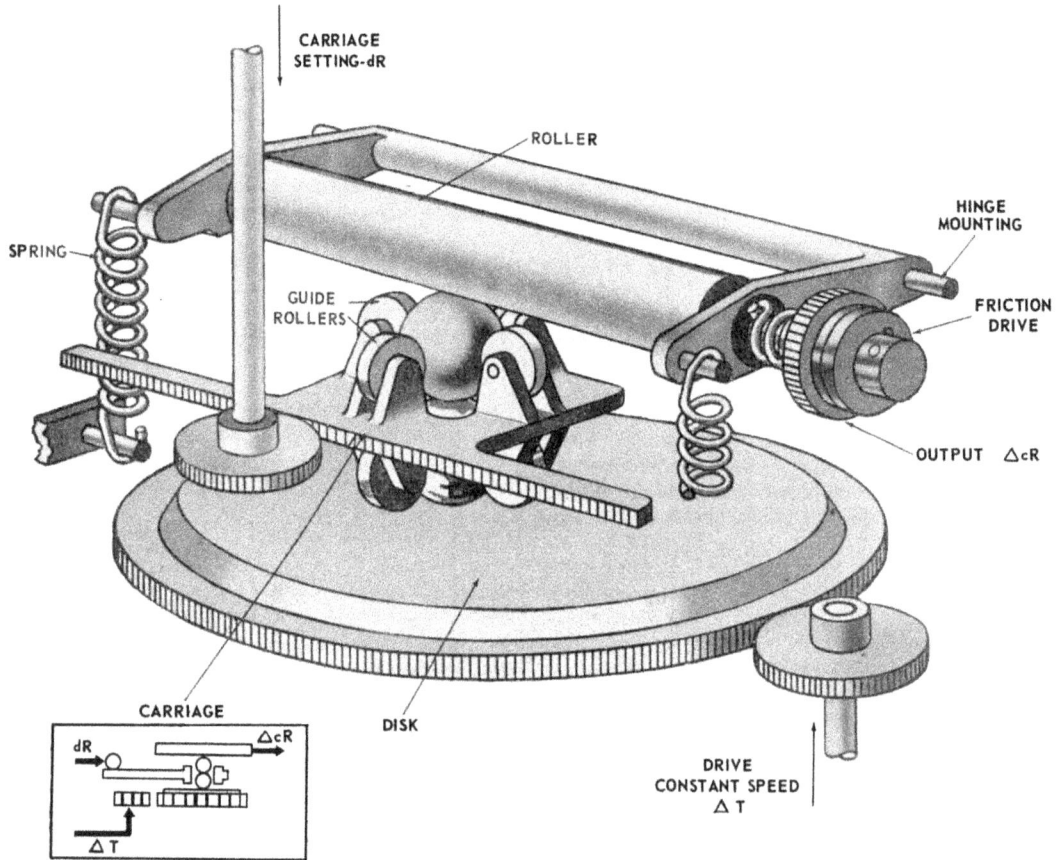

110.12

Figure 15-15.—The disc type integrator.

INDEX

Accumulator, 67
Anchor winch, 36, 38
Automobile valve gear, 36
Axles used in power trains, 148, 149

Balancing moments of force, 18
Ballast tank, 67, 68
Ball bearings, 71, 73
Barometer, 54-57
Bearings, 69-73
Bendix-Weiss universal joint, 79
Bevel gear, 31
Bevel gear differential, 75
Block and tackle, 10-15, 41
    applications afloat and ashore, 13
    chain hoist, 15
    flaghoist, 10
    gun tackle, 11
    luff tackle, 11
    luff upon luff, 13, 14
    mechanical advantage, 10, 15
    runner, 10
    yard and stay tackle, 13
Bourdon gage, 53
Brakes, 28, 63
    friction, 28
    hydraulic, 63
Breech, 16-inch gun, 7

Cams, 85
Cam and cam followers, 80-82
Cam-driven valve, 36
Cam shaft, 36
    of a 4-stroke cycle engine, 127
Capstan bar, 17, 18
Chain hoist, 15
Chain stopper, 8
Clockwise moments of force, 19
Clutches, 82, 86, 130-133
    power train, 130-133
Combustion engines, 106-129
    basic strokes, 108
    classification of, 113

construction of, 115-129
    auxiliary assemblies, 129
    connecting rods, 123
    crankcase, 117
    crankshaft, 124
    cylinder head, 117
    engine cylinder block, 115
    engine head, 126
    exhaust manifold, 118
    gaskets, 119
    intake manifold, 118
    moving parts, 119-129
    piston assembly, 120
    stationary parts, 115-119
    timing gears, 128
    valves and valve mechanisms, 126-128
    vibration damper, 125
cycles, 109-112
development of power, 106
diesel engines, 114
gasoline engines, 114
multiple-cylinder engines, 112
one-cylinder engine, 106-112
Computer mechanisms, 150-157
    differentials, 150-151
    linkages for adding and subtracting,
    150-156
        adding lever, 152
        angle resolver, 156
        component solver, 155
        integrators, 156
        multipliers, 151-156
        multiplying linkage, 155
        screw type multipliers, 152
Constant mesh transmission, 136
Counterclockwise moments of force, 19
Couple, 20
Couplings, 78-84
Crane, electric, 7
Curved lever arms, 5
Cylinders used in power trains, 131

Deep-sea diver, 60
Depth charge, 59

Diesel engine, 106, 114
Differentials, gear, 144-148
Dogs, 7
Drill press, 9

Electric crane, 7
External combustion engine, 106, 107
External mesh gears, 35

Final drives, 142
Flaghoist, 10
Fluids, pressure exerted by, 52
Force and pressure, 50-57
   force, measurement of, 50
   pressure, measurement of, 50-56
      barometers, 54-56
      exerted by fluids, 52
      gages, 53
      manometer, 56
      scales, 50, 51
Four-speed truck transmission, 133
Friction, 40, 42-45
Friction brake, 28
Fulcrum, 17, 19-21

Gages used to measure pressure, 53
Gangplank, 24
Gasoline engine, 106, 114
Gear differential, 74-78, 144-148
   used in automobiles and trucks, 144-148
Gears, 30-38
   anchor winch, 36, 38
   automobile valve gear, 36
   cam-driven valve, 36
   cam shaft, 36
   changing speed, 34
   external mesh gears, 35
   idler gear, 35
   internal gear, 31
   magnifying force with, 35
   pinion gear, 30
   rack and pinion as a steering
     mechanism, 37
   spur gear, 30
   types of, 30-34
   worm gears, 32
Gear trains, 128
Gun tackle, 11

Helical gears, 30
Herringbone gears, 31
Hooke joint, 79
Horsepower, 46-48

Hydraulic machines, 62-68
   advantages of, 63
   brakes, 63
   liquids in motion, 62
   Pascal's law, 62
   press, 64
   steering with, 65
   used on submarines, 66
Hydraulic valve lifter, 127
Hydrostatic machines, 58-61

Idler gear, 35
Inclined plane, 23-26
   applications afloat and ashore, 24
   barrel role, 23
   gangplank, 24
   in spiral form, 26
   ramp, 23
   wedge, 23
Internal combustion engines, 106-129
   basic strokes, 108
   classification of, 113
   construction of, 115-129
     auxiliary assemblies, 129
     connecting rods, 123
     crankcase, 117
     crankshaft, 124
     cylinder head, 117
     engine cylinder block, 115
     engine head, 126
     exhaust manifold, 118
     gaskets, 119
     intake manifold, 118
     moving parts, 119-129
     piston assembly, 120
     stationary parts, 115-119
     timing gears, 128
     valves and valve mechanisms, 126-128
     vibration damper, 125
Internal gear, 31

Jack, 41
Jack screw, 26, 27

Levers, 1-9
   chain stopper, 8
   classes of, 2
   curved lever arm, 5
   dogs, 7
   drill press, 9
   electric crane, 7
   oars, 2
   pelican hook, 8
   wrecking bar, 7

Linkages, 78
Luff tackle, 11

Machine elements and basic mechanisms, 69-86
   basic mechanisms, 74
   bearings, 69-73
     ball bearings, 73
   cams, 80-82, 85
     and cam followers, 80-82
   clutches, 82, 86
   couplings, 78-84
   gear differential, 74-78
   linkages, 78
   radial-thrust roller bearing, 73
   springs, 71-74
   types of, 74
   universal joints, 79-81
Manometer, 56
Micrometer, 27
Moment of force, 17, 20
Multiple-cylinder engines, 112

Oars, 2
Oldham coupling, 78, 80
One-cylinder engine, 106

Pelican hook, 8
Pinion gear, 30
Pointer's handwheel, 21
Power, 46-57
   calculation of, 46-48
   horsepower, 46
   motor, 48
     Prony brake, 48
Power takeoffs, 141
Power trains, 130-149
   axles, 148, 149
   clutch, 130-133
   differentials, 144-148
   final drives, 142-144
   joints, 142, 145
   propeller shaft assemblies, 142
   transfer cases, 140-142
   transmission, 133-139
Pressure and the deep sea
  diver, 60
Pressure, measurement of, 50-56
   balances, 50, 52
   barometers, 54-56

exerted by fluids, 52
   gages, 53
   manometer, 56
   scales, 50, 51
Prony brake, 48

Quadrant davit, 29

Rack and pinion, 37
Radial-thrust roller bearing, 71, 73
Ramp, 23
Rigger's vise, 28
Roller bearing, 71, 73
Roller bitt, 43
Runner, 10

Scales, 50, 51
Schrader gage, 53
Screw applications, 26-29
   friction brake, 28
   inclined plane in spiral form, 26
   jack screw, 26, 27
   mechanical advantage of, 29
   micrometer, 26
   quadrant davit, 29
   rigger's vise, 28
   turnbuckle, 28
Sleeve coupling, 78
Spider gears, 76
Spiral bevel gears, 144
Springs, 71-74
Spur gears, 30
Steam engine, 106
Synchromesh clutch, 138
Synchromesh transmission, 136

Timing gears in engines, 128
Torpedo, 60
Torque, 17, 21, 22
   wrench, 21
Transfer cases, 140-142
Truck transmission, 133
Turnbuckle, 28

Universal joints, 79-81

Valve lifters, 127
Valves and valve mechanisms, engine, 126-128

Watertight door, 88
Wedge, 23
Wheel and axle, 16-25
    balancing moments of force, 18
    capstan bar, 17, 18
    clockwise moments of force, 19
    counterclockwise moments of
     force, 19
    couple, 20
    fulcrum, 17, 19, 20, 21
    mechanical advantage, 16
    pointer's handwheel, 21
    rotation, 17

    torque, 17, 21, 22
    wrench, 21
Work, friction, and efficiency, 39-45
    block and tackle, 41
    friction, 42-45
    jack, 41
    measurement of, 39-41
    roller bitt, 43
Worm and worm wheel, 32
Worm gear, 32
Wrecking bar, 7

Yard and stay tackle, 13